Finn's Thermal Physics

Finn's Thermal Physics

Fourth Edition

Andrew Rex and C.B.P. Finn

CRC Press
Taylor & Francis Group
Boca Raton London New York

CRC Press is an imprint of the
Taylor & Francis Group, an **informa** business

Designed cover image: Shutterstock_ 2117136530

Fourth edition published 2024
by CRC Press
2385 NW Executive Center Drive, Suite 320, Boca Raton FL 33431

and by CRC Press
4 Park Square, Milton Park, Abingdon, Oxon, OX14 4RN

CRC Press is an imprint of Taylor & Francis Group, LLC

© 2024 Andrew Rex

Third edition published by CRC Press 2017

ISBN: 978-1-032-28903-8 (hbk)
ISBN: 978-1-032-28982-3 (pbk)
ISBN: 978-1-003-29947-9 (ebk)

DOI: 10.1201/9781003299479

Typeset in Utopia
by Deanta Global Publishing Services, Chennai, India

Contents in Brief

Contents

Preface to the Fourth Edition

I have been delighted by the feedback I've received on the third edition of *Finn's Thermal Physics*. In particular, the enhanced problem sets and the introduction of two significant chapters on statistical physics seem to have had the positive effects I intended. I appreciate all of you who have sent your comments and suggestions, either to me directly or to the editors.

In preparing this fourth edition, I worked through the entire text to make sure it contained the latest information and relevant updates since I published the third edition about seven years ago. In this process, I considered every comment and suggestion I had received from users. In many places I revised the text accordingly. Sometimes even a small wording change results in a great improvement in clarity and accuracy, and I hope that you find this is the case when you use the fourth edition.

One change that you will notice is the introduction of worked examples in marked boxes. This is a common feature of introductory physics textbooks that is usually absent from more advanced texts like this one. Students often comment on this, and I decided this new edition was the time to insert a few well-chosen worked examples to help the students transition between the text and the problem sets. I'm grateful to the editors for expanding the page limit to allow this new feature.

The people who have helped produce the fourth edition are too numerous to name individually, but there are a few people I want to thank in particular. Of all the user comments I received, the most extensive and useful ones came from Rayf Shiell of Trent University. I also want to recognize the work done by Taylor & Francis editors Carolina Antunes, Betsy Byers, Emma Morley, and Haylie Allan.

Andrew Rex

Preface to the Third Edition

Thermal physics is a beautiful subject that is rooted in the real world but has strong connections to other basic areas of physics—classical dynamics, electromagnetism, and quantum theory—as well as to the disciplines of chemistry and engineering. Everyone has a sense of what happens when they put ice into a drink or open the front door on a cold day. However, the subject is full of subtleties that only emerge upon deeper study. I have found that generally students are happy and grateful to see these, to build on their experience-based intuition, and to gain the expertise that enables them to solve more challenging problems.

Given the beauty and importance of this subject, I was delighted when Luna Han, my editor at Taylor & Francis, asked me to consider working on a revision of Finn's *Thermal Physics*. Moreover, on a personal level, this project dovetails with my own interests and expertise in statistical mechanics. I first encountered the Maxwell's demon problem over 30 years ago, when teaching the brief section on statistical mechanics in a modern physics course (second-year undergraduates). One day, about 15 minutes before class, I thought that I might quote Maxwell's original conception and then explain why a Maxwell's demon can't work. Needless to say, I was unable to come up with the explanation in that timeframe. Here we are now, over 30 years later, and there are still new demons and other challenges to the second law being invented with some regularity. The second law invites such challenges because in its statistical formulation it expresses only strong probabilities, not certainties. (This is just one of those subtleties I mentioned above!) My own thinking is that no such challenge has yet proved sufficient, and to further the discussion I like to challenge people to give me a computer that I can plug into my bathtub. I also remind people what was said on the subject by Arthur Eddington in 1935:

> The law that entropy always increases—the second law of thermodynamics—holds, I think, the supreme position among the laws of Nature. If someone points out to you that your pet theory of the universe is in disagreement with Maxwell's equations—then so much the worse for Maxwell's equations. If it is found to be contradicted

by observation, well, these experimentalists do bungle things sometimes. But if your theory is found to be against the second law of
thermodynamics I can give you no hope; there is nothing for it but to
collapse in deepest humiliation.

Obviously, even today this is somewhat contentious.

In embarking on this revision, I had as a starting point an outstanding text in
the second edition of *Finn's Thermal Physics*. The literature is full of positive
user reviews, and there are many loyal users of this book. Thus, my greatest
challenge has been to add what I could to an already excellent resource without diminishing the effectiveness of the core material. I expect those familiar
with Finn's second edition to find much of this book, even most of it, quite
recognizable.

One notable feature of Finn's book is that it presented such a complete picture of thermodynamics with a fairly minimal inclusion of the approach to
the subject via statistical mechanics. However, this is just where I felt a major
enhancement was in order. The third edition offers two brand-new chapters:
Chapter 6 devoted to classical statistics and Chapter 13 introducing quantum
statistics. These additions are not only useful to the student, but it is also beautiful to see how classical thermodynamics and statistical mechanics lead to
identical results. At the same time, the new chapters and those in between are
designed so anyone who wishes to can skip over some or all of the new material without loss of continuity.

Another enhancement in the third edition is in the problem sets, which are
now placed more prominently and traditionally at the end of each chapter
rather than in an appendix. I have augmented the problems, not only in the
new chapters (6 and 13) but also by adding problems to every chapter, in some
cases roughly doubling the size of the problem set. Many of the new problems
are "battle-tested" in my own classes or exams. Whenever possible, I have
focused on added problems that present practical outcomes and require computation. Similarly, I have added some examples throughout the main narra-
tive as a way of illustrating the theory already so well presented by Finn.

No project of this magnitude is the work of a single individual, and I have many
people to thank for their contributions to the third edition. Luna Han has been
a supportive and resourceful editor at every stage. Several reviewers contrib-
uted a number of useful comments regarding the project as a whole and then
specifically on drafts of the new material, including Carl Michal (University

of British Columbia), Yoonseok Lee (University of Florida), Kevin Donovan (Queen Mary University of London), John Dutcher (University of Guelph), and Steven Bramwell (University College London). In a more global sense, it has been my privilege to work with many colleagues who inspired and enriched my work in thermal and statistical physics. First among these is Harvey Leff (Cal State Poly University, Pomona), my longtime friend and colleague with whom I coauthored two books and several articles. I also want to recognize the work of Daniel Sheehan (University of San Diego), who has organized and hosted a number of important conferences on challenges to the second law. Over the years I have been fortunate to have many great colleagues here at the University of Puget Sound who have enhanced my understanding of this and other subjects. These include Jim Clifford, Fred Slee, Frank Danes, Alan Thorndike, Jim Evans, Greg Elliott, Bernie Bates, Amy Spivey, Randy Worland, Tsunefumi Tanaka, David Latimer, and Rachel Pepper. Then there are the students here at Puget Sound whose intelligence and enthusiasm has made me the best possible teacher and writer. Finally, I have enjoyed the constant support of my family, particularly my wife, Sharon. She has never failed to encourage my work and has often reminded me of its importance in the world.

Andrew Rex

Author

Andrew Rex is a professor of physics at the University of Puget Sound in Tacoma, Washington. He earned a BA in physics at Illinois Wesleyan University in 1977 and a PhD in physics at the University of Virginia in 1982. At Virginia he worked under the direction of Bascom S. Deaver Jr on the development of new superconducting materials. After completing requirements for his PhD, he joined the faculty at Puget Sound. Dr Rex's primary research interest is in the foundations of the second law of thermodynamics. He has published research articles and, jointly with Harvey Leff, two comprehensive monographs on the subject of Maxwell's demon (1990, 2003). Dr Rex has coauthored several widely used textbooks—*Modern Physics for Scientists and Engineers* (1993, 2000, 2006, 2013, 2021), *Integrated Physics and Calculus* (2000), and *Essential College Physics* (2010, 2021)—and the popular science book *Commonly Asked Questions in Physics,* also published by Taylor & Francis/CRC Press. Dr Rex has served in administrative roles, including chair of his department and director of the University of Puget Sound Honors Program. He is devoted to physics education and has been an active participant in the American Association of Physics Teachers, the Society of Physics Students, Sigma Pi Sigma, and Sigma Xi. In 2004, Dr Rex was recognized for his teaching with the President's Award for Teaching Excellence.

Introduction

C.B.P. Finn

THE IMPORTANCE OF THERMODYNAMICS

The science of thermodynamics was developed in the 19th century mainly out of an interest in heat engines—the steam engine and the internal combustion engine. It concerns itself with the relationships between the large-scale bulk properties of a system that are measurable, such as volume, temperature, pressure, elastic moduli, and specific heat. These are often called macroscopic properties. Thus, thermodynamics belongs to classical physics.

Modern physics, on the other hand, attempts to explain the behavior of matter from a microscopic or atomic viewpoint using the techniques of quantum and statistical mechanics. You might ask, then, why we bother with this classical subject of thermodynamics.

The answer is that the modern physics approach of quantum and statistical mechanics depends for its accuracy on the correctness of the microscopic model chosen to represent the physical system. By a microscopic model we mean a *simplified* picture of the system consisting of a collection of small atomic-sized particles. For example, a possible model of a crystal of common salt could be a collection of sodium and chlorine ions alternately placed at the corners of a stack of cubes, the forces between the ions being represented by springs. The accuracy of these models is often dubious, as must be any calculations based on them. Thermodynamics, on the other hand, is not dependent on any such microscopic model and it is important for that very reason. The results of quantum mechanics and statistical mechanics, when scaled up to macroscopic proportions, have to give results consistent with thermodynamics, and so we have an important check on our microscopic picture. However, thermodynamics by itself can give us no fine microscopic details: it can tell us only about the bulk properties of a system.

Classical thermodynamics, then, has a relevance within the framework of modern physics and is as important today as it ever was. This point was brought home by Einstein who in 1949 said:

> A theory is the more impressive the greater the simplicity of its premises, the more varied the kinds of things that it relates and the more extended the area of its applicability. Therefore, classical thermodynamics has made a deep impression upon me. It is the only physical theory of universal content which I am convinced, within the areas of the applicability of its basic concepts, will never be overthrown.

C.B.P. Finn
Second Edition 1993

Chapter 1: Temperature

The concept of temperature is fundamental to any study of thermodynamics. You have an intuitive sense of temperature because you can feel if an object is hot or cold. However, like some other fundamental quantities in physics (think, for example, of time or electric charge), it is not easy to give a precise definition of temperature. To do so, it is necessary to define some other basic concepts and introduce the so-called "zeroth law of thermodynamics," which leads to the definition of thermal equilibrium; from this, temperature can be defined in an unambiguous way.

1.1 BASIC CONCEPTS

In thermodynamics, attention is focused on a particular part of the universe, simply defined as the *system*. The rest of the universe outside the system is called the *surroundings*. The system and the surroundings are separated by a *boundary* or *partition*, and they may, in general, exchange energy and matter, depending on the nature of the partition. For now consider the exchange of energy only, which makes for a *closed system*, so that there is no matter exchange between system and surroundings.

1.1.1 A System, Its Walls, and Surroundings

A useful example of a system is a fixed mass of compressible fluid, such as a gas, contained in a cylinder with a moveable piston as shown in Figure 1.1. This simple system serves as a model for developing some important ideas in thermodynamics.

First, consider a system that is completely isolated from its surroundings. The degree of isolation from external influences can vary over a very wide range,

DOI: 10.1201/9781003299479-1

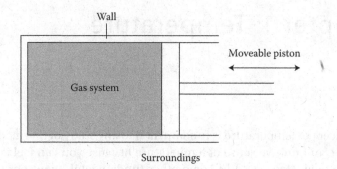

Figure 1.1 A gas contained in a cylinder is a useful example of a system.

and it is possible to imagine walls that make the isolation complete. In practice, the rigid walls of an ordinary vacuum flask are a good approximation to completely isolating walls. It is an important fact of experience that, after a time, this gas system, or any other system contained in such isolating walls, tends to an *equilibrium state* in which no further changes occur. In particular, the pressure P becomes uniform throughout the gas and remains constant in time, as does the volume V. We say that the gas is in an equilibrium state (P, V). It is a further fact of experience that specifying these equilibrium values of the pair of independent variables P and V, together with the mass, fixes all the macroscopic or bulk properties of the gas—for example, the thermal conductivity and the viscosity. A second sample of the same amount of gas with the same equilibrium values for P and V, but not necessarily of the same shape, would have the same viscosity as the first. These ideas can be generalized into the following definition:

> **An equilibrium state is one in which all the bulk physical properties of the system are uniform throughout the system and do not change with time.**

Later you will see that there are other simple thermodynamic systems, apart from a gas, where it is necessary to use other pairs of independent variables to specify the equilibrium state. For a stretched wire system, for example, the appropriate pair is the tension F and length L. Other examples will arise in Section 1.2.3. The important point is that two variables are required to specify the equilibrium state of a simple system. Such directly measurable variables are called *state variables*. Other common names are *thermodynamic variables* and *thermodynamic coordinates*.

1.1.2 State Functions (Properties)

The basic state variables P, V, and temperature T are used to define other functions, which take unique values at each equilibrium state. Some examples of such functions are the internal energy, the entropy, and the enthalpy. They are examples of *state functions*. It is important to realize that it does not matter how a particular state was reached; the value of a state function is always the same for a system in a given state and in no way depends on its past history. *State property* is an alternative and perhaps a more appropriate name for state function. It will be shown in Section 1.3 that P, V, and T are functionally connected at each equilibrium state by the equation of state, and therefore it is possible to express any one in terms of the other two. Thus, these quantities are themselves *state functions*, but we give them the additional name of *state variables* because they are easily measured and allow us to specify an equilibrium state in a convenient, practical way.

1.1.3 Adiathermal (Adiabatic) and Diathermal Walls

There are different ways to change the pressure or volume of a gas system. For example, the piston in Figure 1.1 can be pushed in (to the left). This is an example of a *mechanical interaction* between the system and the surroundings. Pushing the piston in clearly reduces the volume, and in the process the gas's pressure is likely to change too.

Suppose now that no mechanical interaction is allowed to occur—as would be the case if the piston were clamped, with the walls now being rigid, so that the gas's volume is constant. Consider a second cylinder, fitted with a free piston, containing the same gas with the same mass, volume, and pressure as the first. Let the two cylinders be put into contact, as shown in Figure 1.2, and let

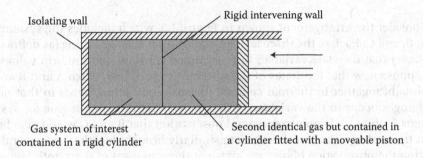

Figure 1.2 An arrangement for determining whether or not a wall is adiabatic.

the piston of the second cylinder be pushed in. Depending on the nature of the intervening wall between the cylinders, there may or may not be changes in the pressure of the gas system in the first cylinder. If there is no change, the intervening wall is said to be *adiathermal* or, more commonly, *adiabatic*; if there is a change, the wall is said to be *diathermal*, and a *thermal interaction* has taken place. A wall made of metal such as copper or aluminum is a good approximation to a diathermal wall, while a good realization of an adiabatic wall is that of a vacuum flask. Two systems in contact via a diathermal wall are said to be in *thermal contact*.

A remark should be made at this point. The reader may wonder why diathermal and adiabatic walls are not defined according to whether or not they conduct heat. The answer is that while such walls have those properties, we cannot address the issue in this way until heat is defined in Chapter 3.

1.2 EQUILIBRIUM STATE

If two thermodynamic systems such as gases are put into thermal contact, after a time no further changes in the pressures and volumes will occur. When the gases' pressures and volumes are no longer changing, each gas is then considered to be in an *equilibrium state*, and the gases are said to be in *thermal equilibrium* with each other, thereby leading to the definition of *temperature*.

1.2.1 Thermal Equilibrium and the Zeroth Law of Thermodynamics

Consider the arrangement shown in Figure 1.3, which includes the systems A, B, and C. Each of the three is in an equilibrium state, meaning (as defined above) that the state variables have assumed constant and uniform values. Suppose now that the states of the systems are such that, when A and B are brought together in thermal contact, thermal equilibrium exists in that no changes occur in the variables. Also suppose that the same is true for systems A and C. It is an experimental observation that B and C would also be in thermal equilibrium if they were similarly brought together. By generalizing the observation above, we arrive at the statement of the *zeroth law of thermodynamics*:

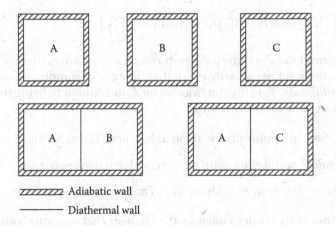

Adiabatic wall

Diathermal wall

Figure 1.3 An illustration of the zeroth law of thermodynamics. If A and B are in thermal equilibrium upon contact, as are A and C, then so are B and C.

> **If each of two systems is in thermal equilibrium with a third, they are in thermal equilibrium with one another.**

1.2.2 Temperature

The preceding experimental observation is the basis of the concept of temperature. It follows from the zeroth law that a whole series of systems could be found that would be in thermal equilibrium with each other were they to be put in thermal contact—a fourth system, D, which is in thermal equilibrium with system C would also be in thermal equilibrium with A and B, and so on. All the systems possess a common property called *temperature*, *T*.

> **The temperature of a system is a property that determines whether or not that system is in thermal equilibrium with other systems. Systems in thermal equilibrium with one another have the same temperature, *T*.**

More formal mathematical arguments may be developed to show the existence of temperature, but they will not be presented here (see, e.g., Adkins 1984; Zemansky and Dittman 1997).

1.2.3 Thermodynamic Equilibrium

If two systems have the same temperature so that they are in thermal equilibrium, this does not necessarily mean that they are in complete or *thermodynamic equilibrium*. For this condition to hold, in addition to being in thermal equilibrium, they also have to be in

1 *Mechanical equilibrium*, with no unbalanced forces acting.

2 *Chemical equilibrium*, with no chemical reactions occurring.

3 *Diffusive equilibrium*, with no flow of matter from one system to another.

Much of thermodynamics concerns the changes that occur to both systems when one or more of these three kinds of equilibrium do not exist. In later chapters they will each be considered in turn.

1.2.4 Isotherms

Consider again the gas system contained in a cylinder with a moveable piston, as in Figure 1.1. Suppose that the gas in the equilibrium state (P, V) and is in thermal equilibrium with another reference system that surrounds the cylinder, so that the two systems have the same temperature. This state can be plotted as a point on a pressure versus volume plot, which is called an *indicator diagram* or *PV diagram*.

Let the gas system be separated from the reference system. If the piston is now pushed in to take the gas to a new state (P', V'), and if this new state is also in thermal equilibrium with the unchanged reference system, then by the zeroth law the two states (P, V) and (P', V') are themselves in thermal equilibrium and have the same temperature. This really means that two identical systems in the states (P, V) and (P', V') would be in thermal equilibrium. The locus of all such points with the same temperature is called an *isotherm*. The isotherms for an ideal gas, to be discussed in the following section, are shown in Figure 1.4.

1.3 EQUATIONS OF STATE

You have seen that all the bulk physical properties of a system in an equilibrium state are fixed by specifying two independent state variables, and these

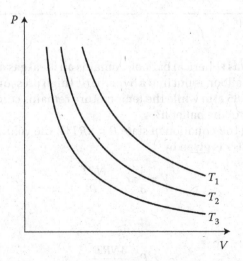

Figure 1.4 The isotherms for an ideal gas. They form a family of hyperbolae.

properties must include the temperature. For a gas, this implies that there is a functional relationship between P, V, and T:

$$f(P,V,T)=0$$

Such a relation is called an *equation of state*. It shows that of the three directly measurable variables, P, V, and T, only two are independent and any one may be expressed in terms of the other two. The state of the gas is equally well specified by quoting (P, V), (P, T), or (V, T).

As an example of an equation of state, consider an ideal gas (where there are no intermolecular attractions and the molecules themselves have no volume) where the equation of state is determined empirically to be

$$PV = nRT$$

Here n is the number of moles present and R is a constant called the universal gas constant, with approximate value $R = 8.314$ J/(mol·K). It follows from this equation of state that the isotherms for an ideal gas shown in Figure 1.4 are a family of hyperbolae, following the equation

$$P = nRT/V$$

Equations of state for systems other than a gas are presented later in this book.

EXAMPLE 1.1

A 25.0-cm-radius spherical balloon contains an ideal gas at a pressure of 1.01 atm. The balloon is put into a hyperbaric (high pressure) chamber at a pressure of 1.85 atm while the temperature remains constant. What is the new radius of the balloon?

Solution: By the equation of state $P = nRT/V$, the volume for the balloon with radius r is given by

$$V = \frac{4}{3}\pi r^3 = \frac{NRT}{P}$$

Rearranging,

$$r^3 P = \frac{3NRT}{4\pi}$$

In this case, all the quantities on the right side of the equation remain constant in this process, and therefore $r^3 P$ = constant. Then the initial quantities (i) are related to final quantities (f) by

$$r_i^3 P_i = r_f^3 P_f \quad \text{or}$$

$$r_f = \left(\frac{P_i}{P_f}\right)^{1/3} r_i = \left(\frac{1.01 \text{ atm}}{1.85 \text{ atm}}\right)^{1/3} (25.0 \text{ cm}) = 20.4 \text{ cm}$$

Under higher pressure, the balloon has been compressed to a smaller volume. That is consistent with the sense of the equation of state, in which V and P are reciprocal quantities.

1.4 SCALES OF TEMPERATURE

From the discussion to this point, it might seem that temperature is a fairly abstract quantity that would be difficult to measure, perhaps requiring a system of cylinders and pistons with one or more reference systems. However, from an early age people develop an intuitive sense of temperatures, warmer or colder, and later learn to associate them with specific numbers. Those numbers appear on thermometers, which are ubiquitous in modern cultures and normally (in the 21st century) come with digital displays. For example, you know that 30°C (or 86°F) is a warm day, while 0°C (or 32°F) is a cold day. You know that your body temperature should be close to 37°C (or 98.6°F).

1.4.1 Absolute Thermodynamic Temperature Scale

In order to give *numerical* values to different temperatures, a systematic and reproducible method for assigning such values is required. The first task is to choose a system and then to select a physical property of that system (the *thermometric variable* or *thermometric property*) that varies with temperature. In order to make the argument general, the thermometric variable will be labeled X.

The normal choice of X is something that can easily be measured, such as the length of the column of mercury in a mercury-in-glass thermometer or the resistance of a piece of platinum wire. Unfortunately, scales of temperature defined using different but familiar thermometric variables do *not* on the whole agree throughout a wide range of temperatures, although in practice the differences are small. This point will be addressed in more detail later.

The general thermometric variable X is used to set up a scale of temperature. Call T_X the temperature on the X scale, where the subscript X is a reminder that the temperature depends on the thermometric property chosen. The numerical value of temperature on this scale is defined so that the thermometric property X varies with temperature in the simplest possible way, according to the linear relation

$$X = cT_X \qquad \qquad 1.1$$

where c is a constant. The value of c is fixed by choosing an easily reproducible T_X (*a fixed point*) and assigning to it a *particular* value. The customarily chosen fixed point is the temperature at which ice, water, and water vapor coexist in equilibrium; this is known as the *triple point of water*. The value given to T_X at this fixed point is 273.16—the choice of the value 273.16 will be discussed shortly. Substituting this value for the temperature of the triple point, where the value of X is X_{TP}, in Equation 1.1 gives

$$T_X = 273.16(X/X_{TP}) \qquad \qquad 1.2$$

There are two issues here that deserve discussion:

1 Equation 1.2 implies a zero of temperature on the X scale, that is, $T_X = 0$, when $X = 0$. In practice, such a T_X may not occur if the thermometric variable does not vanish as the temperature is progressively lowered. For example, the resistance of a length of platinum wire always remains non-zero, becoming non-linear and tending to a constant value at the lowest attainable temperatures. The ideal gas scale, to be discussed below, does have a meaningful zero of temperature because the thermometric property used there, the pressure, eventually vanishes as the temperature is extrapolated to zero.

2 Temperatures on the X scale are defined only in regions where Equation 1.2 is meaningful. If, for example, you are using a mercury-in-glass thermometer with X being the length of the column, this equation gives a temperature only as long as there is a measurable length of mercury in the capillary. At low temperatures, when the mercury has dropped back into the bulb, Equation 1.2 has no relevance. This is one reason why, in practice, the mercury-in-glass scale defined according to Equation 1.2, is not used, even though in principle such a scale is possible. Instead, these thermometers are calibrated in terms of other standard ones such as those described in Section 1.4.5.

Note that before 1954 temperature scales were based on the modified relation

$$X = cT_X + d \qquad\qquad 1.3$$

Then the two constants c and d had to be fixed by specifying the temperature at *two* fixed points, the steam and the ice points, which are the temperatures of boiling and freezing water at 1 atm pressure. Since 1954, Equation 1.2 has been used, requiring only *one* fixed point. See Section 1.4.5 for further discussion of this point.

1.4.2 Limitations of the Thermodynamic Scale

It is important to realize that different thermometers based on different thermometric variables will agree by definition only at fixed points. At other temperatures a mercury-in-glass thermometer will give slightly different values for a particular temperature than, say, a resistance thermometer, which correlates the resistance of the detection element (such as platinum, nickel, or copper) with temperature. The resistance of any metal or alloy does not vary in a linear way with temperature, and the amount of deviation from linearity depends on the temperature range being considered. Therefore, the linear relationship expressed in Equation 1.1 is only an approximation. If the mercury scale and the resistance scale were truly linear, as is suggested by Equation 1.1, then the two scales would agree at all points. However, each of the two thermometers will, in practice, vary from the linear relationship in different ways. Therefore, as the temperature on any thermometer deviates more from its fixed point, a larger deviation from linearity might be expected. Problem 1.6 at the end of this chapter illustrates this point.

Fortunately, there is a class of thermometers that always agree at all points on the temperature scale—gas thermometers. It will be shown in Chapter 4 that temperature defined according to the ideal gas scale has a fundamental significance in thermodynamics and in fact is identical to the temperature, T, on the absolute thermodynamic temperature scale. The development of all our thermodynamic relations will be in terms of T, with the understanding that it can be measured experimentally using a gas thermometer.

1.4.3 The Gas Scale

A schematic diagram of a constant-volume gas thermometer is shown in Figure 1.5. The volume of the gas is kept constant by adjusting the height of the mercury column until the mercury meniscus just touches the marker at the end of the capillary tube. The bulb of gas is immersed in a system whose temperature is to be measured, and the pressure of the gas is used as the thermometric parameter. Allowance has to be made for the fact that some of the gas in the "dead space" may be at a different temperature from that in the bulb. When this and other corrections (not discussed here) have been made, the gas scale temperature is determined from

$$T_{gas} = 273.16(P/P_{TP}) \hspace{3cm} 1.4$$

The number 273.16 in Equation 1.4 is taken to be exact, not rounded. The interesting point is that when the amount of working gas is reduced to as small as possible for measurements still to be made, all gas thermometers give the same temperature for a given system, irrespective of the gas used. Figure 1.6

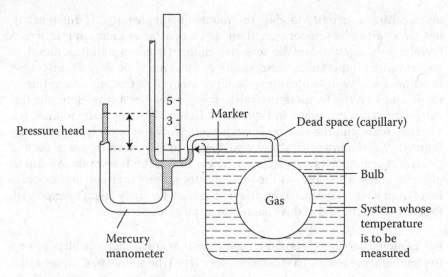

Figure 1.5   A constant-volume gas thermometer.

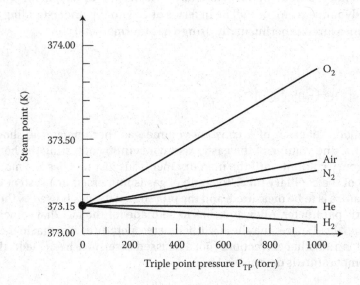

Figure 1.6   The recorded temperature of the steam point as a function of the mass of gas used in a constant-volume gas thermometer.

illustrates this point for the temperature of water boiling under an external pressure of 1 atm, where the limiting value is 373.15. The ordinate in this figure (triple point pressure) is directly related to the amount of gas used, because as the amount of gas is reduced, the corresponding pressure at the triple point is also reduced.

To summarize these findings, the ideal gas scale is defined as

$$T = 273.16 \lim_{P_{TP} \to 0} \left(P/P_{TP} \right) \qquad\qquad 1.5$$

The SI unit for absolute temperature is kelvin (symbol K). Notice that no ° sign is written before K, as it is with other temperature units such as °C and °F. The connection with an ideal gas is that, as the amount of working gas is reduced, the gas becomes closer to an ideal one, as both the intermolecular attraction and the molecular density are reduced.

The apparently curious choice of 273.16 K for the temperature of the fixed point can now be understood. This value was chosen to make the size of the kelvin such that there would be exactly 100 K between the *experimentally determined* temperatures of the ice and steam points on the ideal gas scale. To the accuracy then available, these temperatures were measured to be 273.15 and 373.15 K, respectively. The reader should note the 0.01 K difference between the ice and triple points.

Actually, much to the embarrassment of those who chose the value of 273.16 K for the temperature of the single fixed point, modern measurements have shown that the ice and the steam points differ by slightly less than 100 K! These measurements give a slightly lower value for the steam point at 373.124 ± 0.001 K, so the steam and the ice points differ by only 99.974 ± 0.001 K rather than the intended 100 K. (See, e.g., Pellicer et al. 1999.) This difference is important only for precision work and in general we shall ignore it, noting simply that it exists.

Finally, it should be remarked that the ideal gas scale can equally be defined using constant-pressure thermometers rather than constant-volume thermometers. In practice, only the constant-volume type is used, which justifies the focus of this section.

1.4.4 The Celsius Scale and the International Temperature Scale of 1990

For convenience, it would be desirable to have a temperature scale in which the zero is in the range of commonly encountered temperatures. The Celsius scale t is measured in °C and is defined from to the ideal gas scale T by

$$t(°C) = T(K) - 273.15 \qquad\qquad 1.6$$

with the size of the °C the same as the kelvin. It follows from Equation 1.6 that, ignoring the small recently measured difference from the experimental value of 373.15 K for the steam point as noted in the previous section, the ice point is at 0°C and the steam point is essentially at 100°C. By definition, the temperature of the fixed point, the triple point of water, is 0.01°C (=273.16 K).

Because gas thermometers are cumbersome devices to use, it is convenient to calibrate a whole series of secondary thermometers in terms of the gas scale and to use these where possible. The most recent International Temperature Scale of 1990 (ITS-90) extends upward from 0.65 K to the highest temperature practically measurable in terms of the Planck radiation law using monochromatic radiation. It uses, for example, a platinum resistance thermometer between 14 and 962 K, calibrated at a specified set of fixed points and using a specified interpolation procedure for intermediate temperatures. Below 14 K, a helium gas thermometer is used, except at the lowest temperatures where the temperature is determined from the vapor pressure–temperature relations for ^{3}He and ^{4}He.

1.4.5 Single-Point Temperature Scale

In Section 1.4.1, it was noted that the method of setting the temperature scale was changed in 1954 from a two-point method to the current single-point method. In the old *centigrade* scale (as Celsius was called prior to its renaming in the 9th General Conference of Weights and Measures in 1948), the temperature of the upper fixed steam point was chosen to be 100° centigrade while that of the ice point 0° centigrade. Thus, there should be 100° centigrade difference between these two points for any thermometric variable X by definition— hence the name centigrade. Contrast this with the Celsius scale, defined by Equation 1.6 and set up using a gas thermometer only, where this difference is 100° Celsius by experimental measurement. It follows from Equation 1.3 that, if X_{steam} and X_{ice} are the values of the thermometric variable at the steam and ice points,

$$X_{steam} = 100c + d \qquad X_{ice} = d$$

Substituting the values of the constants c and d obtained from these equations back into Equation 1.3, the centigrade value of the temperature is

$$\theta_x (^\circ\text{centigrade}) = 100 \left[\frac{X - X_{ice}}{X_{steam} - X_{ice}} \right] \qquad 1.7$$

at a general point where the thermometric variable is X. The symbol θ_x is used here to denote the temperature in degrees centigrade, whereas we have used the symbol t earlier to denote the temperature in degrees Celsius. The subscript x on θ is to remind us that, other than at the two fixed points, the centigrade temperature depends on the choice of the thermometric variable. More strictly, one should quote for example a temperature as 20° centigrade (measured on the mercury-in-glass scale) with the words in parentheses added. Note that units ° centigrade are used here, so as to reserve the symbol °C for degrees Celsius.

Suppose the centigrade scale is set up using a gas thermometer, with the results extrapolated to having a vanishingly small amount of gas in the bulb, thus simulating an ideal gas thermometer as in Equation 1.5. Then this gas centigrade scale coincides with the Celsius scale (providing one takes there to be 100 measured kelvins between the steam and ice points and not the more accurate value of 99.974 K as discussed in Section 1.4.3). However, it must be remembered that the two scales are different in principle: in the centigrade scale the steam and the ice points are defined to differ by 100° centigrade; in the Celsius scale they are measured to differ by 100 K, which is the same as 100°C. For precision work the two scales have to be taken as numerically different, with the degree centigrade being slightly smaller than the degree Celsius (100° centigrade as opposed to 99.974° Celsius between the ice and the steam points). You may have heard that Celsius is just a new name for centigrade. This is true only for a gas thermometer within the limitations just discussed. For other thermometers, for example a mercury thermometer, this statement is untrue because the Celsius scale is defined only for a gas thermometer according to Equation 1.6 and has no meaning for other types of thermometers. Such thermometers may, however, read °C if they have been calibrated against an ideal gas thermometer giving °C directly.

Although there are differences between centigrade thermometers using different thermometric variables, the differences are smaller than with the current method utilizing a single fixed point, where you have seen that differences can be quite marked. This does not mean that the current method is inferior to the old method, because these differences are of no importance in practice; all real thermometers are calibrated against the ideal gas scale. The modern method has the enormous advantage of requiring only one fixed point, which halves the problem of standardization between different laboratories.

Also, the old method suffered from a severe disadvantage, so that any uncertainty in the experimental measurement of the thermometric variable X at the two fixed points gives rise to a proportionately larger error when these measurements are extrapolated back to low temperatures. To understand this last point, consider the application of Equation 1.7 to a constant-volume gas thermometer with the pressure P as the thermometric variable:

$$\theta = 100 \left[\frac{P - P_{ice}}{P_{steam} - P_{ice}} \right] °\text{centigrade} \qquad\qquad 1.8$$

It is necessary to measure experimentally the values of P_{steam} and P_{ice} at the steam and ice points which are *defined* as 100° centigrade and 0° centigrade. The results would be as on the left of Figure 1.7.

There are particular difficulties with setting up the ice point (ice coexisting with water-saturated air at 1 atm) in a reproducible way. One problem is that as the ice melts, it tends to surround itself with pure water, insulating it from the water-saturated air. Hence, when the temperature of the ice point is measured with the gas thermometer, there are significant variations in the measured

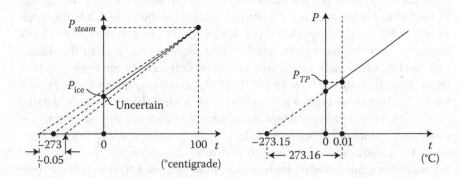

Figure 1.7 Values of P_{ice} and P_{steam} at the steam and ice points.

values of P_{ice}. There are also difficulties in measuring P_{steam} precisely. This is of little importance at ordinary temperatures, but when these measurements are extrapolated back to zero gas pressure to give an intercept of about –273° centigrade, there is a corresponding uncertainty in this intercept of perhaps as much as 0.05° centigrade. If the origin is now taken to be this intercept, that is, absolute zero, then any uncertainty in its value gives a relatively large error at low temperatures. For example, with this quoted uncertainty in the intercept, the boiling point of helium at 4.22 K has an uncertainty of 0.05 K, or about 1%!

This uncertainty explains why the single-fixed-point scale was adopted in 1954. The triple point of water is precisely reproducible, and absolute zero is precisely determined as the limiting temperature at which the pressure in an ideal gas thermometer tends to zero (shown on the right of Figure 1.7). Note that (1) the triple point is 273.16 K above absolute zero by definition; (2) the ice point at 273.15 K is 0.01 K colder than the triple point by measurement; and (3) Equation 1.6 is used as the definition of the supplementary Celsius scale to obtain the values for the temperatures in °C as shown.

Problems

1.1   On the Fahrenheit temperature scale, the ice point and steam point are 32°F and 212°F, respectively. (a) At what temperature do the Fahrenheit and Celsius scales give the same temperature? (b) Find absolute zero (–273.15°C) on the Fahrenheit scale to two decimal places. (c) The temperature you found in (b) is the basis for the Rankine temperature scale, an absolute temperature using Fahrenheit-sized degrees. Express the ice point and steam point in Rankine.

1.2   Consider a gas contained in the cylinder-piston arrangement shown in Figure 1.1. The gas is pure nitrogen (N_2) with a mass of 1.60 g and is in equilibrium with its surroundings. The gas is initially at *standard temperature and pressure*, with $P = 101$ kPa and $T = 0°C$. (a) Find the gas's volume. (b) The piston is now moved, slowly enough so that equilibrium with the surroundings is maintained, until its pressure reaches 120 kPa. What is the gas's new volume?

1.3   The length of the mercury column in a mercury-in-glass thermometer is 5.0 cm when the bulb is immersed in water at its triple point. What is the temperature on the mercury-in-glass scale when the length of the column is 6.0 cm? What will the length of the column be when the bulb is immersed in a liquid at 100° above the ice point, as measured on the mercury-in-glass scale? If the length of the column can be measured to within only 0.01 cm, can this thermometer be used to distinguish between the ice point and the triple point of water? You may take the temperature of the ice point, as measured on the mercury-in-glass scale, as 273.15 K.

1.4   The resistance of a wire is given by

$$R = R_0\left(1 + \alpha t + \beta t^2\right)$$

where t is the temperature in degrees Celsius measured on the ideal gas scale and so R_0 is the resistance at the ice point. The constants α and β are 3.8×10^{-3} K^{-1} and -3.0×10^{-6} K^{-2}, respectively. Calculate the temperature on the resistance scale at a temperature of 70°C on the ideal gas scale.

1.5   The table below lists the observed values of the pressure P of a gas in a constant-volume gas thermometer at an unknown temperature and at the triple point of water as the mass of gas used is reduced.

P_{TP} (torr)	100	200	300	400
P (torr)	127.9	256.5	385.8	516

By considering the limit $P_{TP \to 0}$ (P/P_{TP}) determine T to two decimal places. What is this in °C? (1 torr is a pressure of 1 mm of Hg).

1.6   Different thermometers disagree, except at the fixed points by definition. When using the modern definition of temperature using a single fixed point at the triple point of water (273.16 K) with

$$T_x = T_{TP}\frac{X}{X_{TP}}$$

these differences can be quite significant. To see how big these differences can be, complete the table below for the temperatures T_x of the boiling points of various liquids (at an external pressure of 1 atm) and the melting point of one solid, using thermometers with different thermometric variables X. The values of X_{TP} for the different thermometers are given in the bottom row. Quote your values for the temperature to the nearest degree only. The temperature given for Sn is for the melting point.

	Copper Nickel Thermocouple		Platinum Resistance Thermometer		Constant-Volume H$_2$ Thermometer		Constant-Volume H$_2$ Thermometer	
	E (mV)	T_E	R (Ω)	T_R	P (atm)	T_P	P (atm)	T_P
Liquid/Solid								
N$_2$	−0.10	?	1.96	?	1.82	?	0.29	?
O$_2$	0.00	?	2.50	?	2.13	?	0.33	?
H$_2$O	5.30	?	13.65	?	9.30	?	1.37	?
Sn	9.02	?	18.56	?	12.70	?	1.85	?
At T.P.	2.98	273	9.83	273	6.80	273	1.00	273

Comment on the temperature values for the two constant-volume thermometers. Which one gives values closer to the ideal gas scale?

REFERENCES

Adkins, C.J., *Equilibrium Thermodynamics*, third edition, Cambridge University Press, Cambridge, 1984.

Pellicer, J., Amparo Gilabert, M., and Lopez-Bazea, E., The evolution of the Celsius and Kelvin temperature scales and the state of the art, *Journal of Chemical Education* 76, 911–913, 1999.

Zemansky, M.W. and Dittman, R.H., *Heat and Thermodynamics*, seventh edition, McGraw-Hill, New York, 1997.

Chapter 2: Reversible Processes and Work

Thermodynamics is concerned with changes in the different state functions that occur when a system changes from one equilibrium state to another.

> A *process* is the mechanism of bringing about such a change. These initial and final equilibrium states are called the *end points* of the process.

Pushing in a piston (as in Figure 1.1) and compressing the gas in a cylinder from an equilibrium state (P_1, V_1) to a new equilibrium state (P_2, V_2) is an example of a process.

2.1 REVERSIBLE PROCESSES

There is a particular class of idealized processes that has enormous value in thermodynamics—processes that are *reversible*. They are valuable because it is possible to calculate changes in the state functions for any process using them. This point will be made clear in Section 2.1.4, when considering an example of the thermodynamic method. First, it is necessary to define reversible processes and how they are realized.

Clearly, reversible implies that, in any such change, the system must be capable of being returned to its original state. However, reversible means much more than this in that when the system is returned to its original state, the surroundings must be unchanged too.

A clue to the conditions for reversibility can be gained by considering a pendulum being displaced from one equilibrium position to another by a force

DOI: 10.1201/9781003299479-2

F, as in Figure 2.1. Any force acting through a displacement does work. Let the force $F(\theta)$ be only infinitesimally greater than the restoring force $mg \sin \theta$ at every stage of the displacement, from θ_1 to θ_2. The pendulum then goes through a series of equilibrium states because the process could be stopped at any stage, and the pendulum held where it is. Now if *F* is reduced by a very small amount, the pendulum will move back the other way, with the work done against the force being exactly equal to the work done in the initial displacement. The net work done is zero, and the entire system and its surroundings have been returned to their initial state, which satisfies the conditions for reversibility. This is true provided there are no frictional forces present, such as if the bob moved in a viscous medium. This leads to the following generalizations:

A process that can be thought of as a succession of equilibrium states is called a *quasistatic process*. *Reversible processes* are quasistatic processes where no dissipative forces such as friction are present.

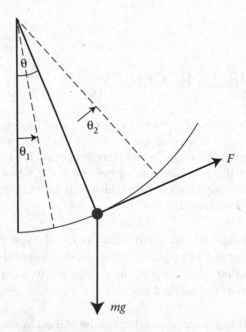

Figure 2.1 A simple pendulum. When the pendulum is displaced, it can be moved through a series of equilibrium states.

2.1.1 Isothermal Compression

For a more relevant thermodynamic example of a reversible process, consider again the gas-cylinder system (Figure 1.1). Let the cylinder undergo an isothermal, reversible compression at temperature T, from the state (P_1, V_1) to the state (P_2, V_2). This could be achieved in the following way. The cylinder is fitted with a frictionless piston and contains the gas in the initial state (P_1, V_1). A force

$$F = P_1 A$$

is applied to the piston to oppose the gas pressure, where A is the area of the piston. The walls of the cylinder are diathermal, and the surroundings at temperature T are so large that this temperature is unaffected by anything done to the gas-cylinder system. Such surroundings are called a *thermal* or *heat reservoir*, or simply a *reservoir*. The external force F is now increased infinitesimally, and the system is allowed to come to a new equilibrium state at the same temperature. This process is repeated until the final state (P_2, V_2) is reached.

It is useful to represent the process on a graph of pressure versus volume, generally referred to as a *PV diagram*, such as the one shown in Figure 2.2. Because the system is always in an equilibrium state, with a well-defined P, V, and T, the process may be plotted as a succession of points (P, V) that forms a continuous curve between the end points (P_1, V_1) and (P_2, V_2). The curve shown in Figure 2.2 is an isotherm, as described in Section 1.2.4, because it represents a process at constant temperature. If the gas is ideal, then the equation of state $PV = nRT$ holds for each point in the process.

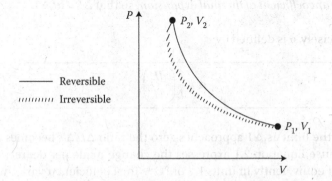

Figure 2.2   A reversible process may be shown as a continuous line on a *PV* diagram. The one shown here is an isothermal process. An irreversible process cannot be shown in this way.

In contrast, it is possible to bring about the same change between the same two end points by pushing in the piston suddenly from volume V_1 to volume V_2. Then there would be turbulence, with finite temperature and pressure gradients both within the gas and between it and the surroundings. Although the gas would eventually settle down to the equilibrium state (P_2, V_2), it does not pass through intermediate equilibrium states, and the process is irreversible. It is impossible to plot this irreversible process on a PV diagram, because the intermediate stages do not have well-defined pressures or temperatures. The irreversible process is represented schematically by a hatched line, as shown in Figure 2.2. Additionally, an equation of state (such as $PV = nRT$) does not hold for the intermediate stages in an irreversible process.

In a reversible process, there are never finite pressure or temperature differences, either within the system or between the system and the surroundings. Further, the direction of a reversible process can be changed by an infinitesimal change in the external conditions.

2.1.2 Thermal Expansion

Solids and liquids generally expand when their temperatures increase. (Gases will expand as much as their container permits, so thermal expansion in gases is harder to quantify.) Consider first a solid metal bar of length L, initially at temperature T. For a small increase in temperature ΔT, the increase in length ΔL is proportional to both L and ΔT. The constant of proportionality is called α, the *linear coefficient of thermal expansion*, so that $\Delta L = \alpha L \Delta T$.

More precisely, α is defined by

$$\alpha = \frac{1}{L}\left(\frac{dL}{dT}\right) \qquad\qquad 2.1$$

where in the limit as ΔT approaches zero the ratio $\Delta L/\Delta T$ becomes a derivative. Because Equation 2.1 expresses the change made per degree, α can be expressed equivalently in units K^{-1} or $°C^{-1}$. The coefficient α varies with temperature, though for many materials it varies slowly over fairly wide temperature ranges. Copper, a fairly typical metal, has $\alpha = 1.65 \times 10^{-5} \ K^{-1}$ at 20°C and $1.83 \times 10^{-5} \ K^{-1}$ at 230°C.

Similarly, in three dimensions the change in volume ΔV of a solid or liquid is proportional to the original volume V and the temperature change ΔT. Now the proportionality constant is β (the *volume coefficient of thermal expansion*), so that $\Delta V = \beta\, V\Delta T$. In the limit as ΔT approaches zero,

$$\beta = \frac{1}{V}\left(\frac{dV}{dT}\right) \qquad\qquad 2.2$$

Thermal expansion is an example of a reversible process, because cooling a substance by an amount ΔT results in a volume decrease equal to the volume increase that occurs when the temperature is increased by ΔT. The volume (or density) of a solid or liquid depends on pressure as well as temperature, so measurements of α or β should be taken at constant pressure. For a solid material, α and β are related (see Problem 2.2):

$$\beta \approx 3\alpha.$$

Linear and volume expansion coefficients for a few sample materials are presented in Table 2.1. More extensive and precise data tables can be found on many web sites. Generally β tends to be larger for liquids than solids, reflecting the fact that liquids expand faster than solids for a given temperature rise. An important exception to this is liquid water just above its freezing point of

TABLE 2.1 Thermal Expansion Coefficients (at $T = 20°C$ Unless Otherwise Noted)

Material	Coefficient of Linear Expansion α $(\times10^{-5}\ °C^{-1})$	Coefficient of Volume Expansion β $(\times10^{-5}\ °C^{-1})$
Solids		
Aluminum	2.4	7.2
Brass	2.0	6.0
Copper	1.7	5.0
Concrete	1.2	3.6
Glass (Pyrex)	3.3	9.9
Lead	2.9	8.7
Silver	1.9	5.7
Steel (typical)	1.2	3.6
Liquids		
Ethanol		75
Mercury		49
Water (1°C)		−4.8
Water (20°C)		21
Water (50°C)		50

0°C at atmospheric pressure. The volume coefficient β decreases as the temperature drops from room temperature and reaches $\beta = 0°C^{-1}$ at $T = 4°C$. From 4°C to 0°C, β is actually negative, which means that the cooling water expands slightly.

EXAMPLE 2.1

Estimate the variation in length of a 3.5-m-long steel beam used in building construction when the temperature varies from 0°C to 30°C.

Solution: As discussed in the text, there should be little variation in the thermal expansion coefficient over that temperature range, so we can use the typical value $\alpha = 1.2 \times 10^{-5} °C^{-1}$ from Table 2.1. Assuming then that α is constant over that range, Equation 2.1 can be approximated by

$$\alpha = \frac{1}{L}\frac{\Delta L}{\Delta T}$$

Rearranging and inserting numerical values,

$$\Delta L = \alpha L \Delta T = \left(1.2 \times 10^{-5} \, °C^{-1}\right)\left(3.5 \text{ m}\right)\left(30°C\right) = 1.26 \text{ mm}.$$

This may seem like a minor amount of expansion, but engineers must consider it in building construction. Thermal expansion of concrete roads is also an issue for engineers. Expansion joints can be inserted in buildings and roads to compensate.

2.1.3 Bulk Modulus

Moduli of elasticity are always given as stress/strain or force/unit area divided by the fractional deformation. For a solid or fluid, *the bulk modulus B* is given by

$$B = -V\left(\frac{\partial P}{\partial V}\right)_T \equiv \frac{1}{\kappa} \tag{2.3}$$

The constant T denotes that the bulk modulus is determined at constant temperature, and so this is the isothermal bulk modulus. The negative sign ensures that B is a positive number because, for all known substances, ΔV is negative for a positive increase in pressure ΔP. The reciprocal of the bulk modulus is the *compressibility κ*.

For a stretched wire of cross-sectional area A, the appropriate modulus of elasticity is *Young's modulus*

$$Y = \frac{L}{A}\left(\frac{\partial F}{\partial L}\right)_T \qquad\qquad 2.4$$

(Note that a positive force ΔF results in an increased length ΔL, so no negative sign is needed in Equation 2.4.) The units for Young's modulus Y are the same as pressure, with SI unit Pa or N/m².

2.1.4 An Example: The Effect of Temperature on Tension in a Wire

As an example of how Young's modulus can be useful in thermodynamics, consider the wire shown in Figure 2.3, clamped between two rigid supports and thus held at constant length L. Suppose the wire is now cooled from T_1 to T_2. Because the wire is not allowed to shrink as it normally would with cooling, the result is an increase ΔF in the wire's tension, which can be found as a function of ΔT.

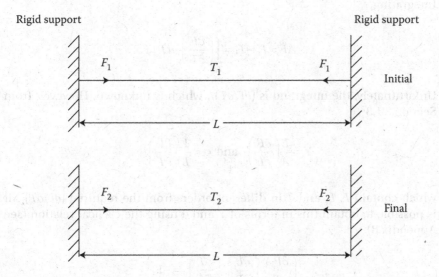

Figure 2.3   A stretched wire being cooled under conditions of constant length.

First, note that the equilibrium states of the wire are fixed by specifying two of the state variables F, L, and T, which are related by some equation of state

$$g(F,L,T)=0$$

The wire undergoes a process in which it is changed from one equilibrium state (F_1, T_1) to another (F_2, T_2), both with the same length.

Suppose for a moment that the wire is cooled from T_1 to T_2 *reversibly*. This could be achieved by bringing up to the wire a whole series of large bodies ranging in temperature from T_1 to T_2 to effect a quasistatic cooling through a sequence of equilibrium states.

For any one of these states $g(F, L, T) = 0$. Or, solving for F,

$$F = F(L,T)$$

where $F(L, T)$ is a function of L and T alone. Then

$$dF = \left(\frac{\partial F}{\partial T}\right)_L dT + \left(\frac{\partial F}{\partial L}\right)_T dL$$

where the second term is zero because the cooling takes place under conditions of constant length.

Integrating,

$$\Delta F = F_2 - F_1 = \int_{T_1}^{T_2} \left(\frac{\partial F}{\partial T}\right)_L dT$$

Unfortunately, the integrand is $(\partial F/\partial T)_L$, which is unknown. However, from Section 2.1.3

$$Y = \frac{L}{A}\left(\frac{\partial F}{\partial L}\right)_T \quad \text{and} \quad \alpha = \frac{1}{L}\left(\frac{\partial L}{\partial T}\right)_F$$

which contain F, L, and T in different orders from the required $(\partial F/\partial T)_L$. It is possible to obtain this in terms of Y and α using the cyclical relation (see Appendix B):

$$\left(\frac{\partial F}{\partial T}\right)_L \left(\frac{\partial L}{\partial F}\right)_T \left(\frac{\partial T}{\partial L}\right)_F = -1$$

therefore,

$$\left(\frac{\partial F}{\partial T}\right)_L = -\left(\frac{\partial F}{\partial L}\right)_T \left(\frac{\partial L}{\partial T}\right)_F = -YA\alpha$$

Finally,

$$\Delta F = -\int_{T_1}^{T_2} YA\alpha dT = -YA\alpha \int_{T_1}^{T_2} dT$$

and

$$\Delta F = -YA\alpha\left(T_2 - T_1\right)$$

if Y, A, and α are independent of T. The result ΔF is positive if $T_2 < T_1$, so the tension in the wire increases as expected.

One might object by saying: This is fine if the cooling is *actually* reversible, but in practice the cooling will not be reversible, because one may simply heat the wire and let it cool. This will result in large temperature gradients both within the wire itself and between the wire and the surroundings. Then the intermediate states are not equilibrium states, and so one cannot apply the equation of state $F = F(L, T)$, and the analysis appears to be invalid.

The answer to this critique is that the preceding analysis is still sound because the wire is being taken between *equilibrium states*. The initial tension of the wire is completely fixed by specifying the initial state (L, T_1) as is the final tension by specifying the final state (L, T_2). Thus the change in the tension is determined by specifying the end points:

$$\Delta F = F_2 - F_2 = F\left(L, T_2\right) - F\left(L, T_1\right)$$

In other words, the change in the state function F does not depend on the path taken from state 1 to state 2. Therefore ΔF is *path independent*, being determined *only by the end points*. It is wise to choose the most convenient path, which happens to be a *reversible* path.

> **The elegant trick just shown is used frequently in thermodynamics to calculate changes in state functions for processes between a pair of equilibrium states.**

2.2 WORK

The remainder of this chapter concerns topics involving work in thermodynamic processes, in preparation for the introduction of the first law of thermodynamics in Chapter 3. A good model for understanding work is the familiar gas cylinder system (Figure 1.1), and ideas developed from this model can be generalized to other systems.

2.2.1 Work in Reversible Processes

Suppose that a gas in the initial equilibrium state (P_1, V_1) is compressed to a new equilibrium state (P_2, V_2) by increasing the external force on the piston and allowing it to slide in. If friction is present, some of the work done on the gas by the piston is expended against these frictional forces. However, if no frictional forces are present, all of this work goes into performing work on the gas. Then it is possible to find a simple expression for the work done in terms of the state variables of the gas, provided that the compression is performed quasistatically, so that the pressure is well defined and uniform throughout the gas. In other words, the gas must be compressed reversibly.

Suppose that at one of the intermediate equilibrium states during the reversible compression, the pressure is P and the balancing force on the piston is F. Then, as shown in Figure 2.4

$$F = PA$$

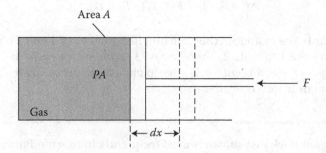

Figure 2.4   The work done on a gas when it is compressed reversibly through an infinitesimal volume change dV is $-P\,dV$.

where A is the area of the piston. If the force is increased infinitesimally so that the piston moves in by dx, the work done on the gas by the surroundings applying the force F is

$$dW = PAdx = -PdV \text{ (reversible)}$$ 2.5

Note that the minus sign in Equation 2.5 is necessary because during compression the volume is decreased, but the work done on the gas is positive. The total work performed on the gas in the process is

$$W = -\int_{V_1}^{V_2} PdV \text{ (reversible)}$$ 2.6

where, as indicated, these results are true for reversible processes.

This reversible work is in fact the maximum work that can be done in a compression or expansion, which can be seen as follows. The work done by compressing the gas is $W = -\int_{V_1}^{V_2} Fdx$ and will be at a maximum when F is as large as possible at all stages in the process. If the gas is to be compressed, the largest possible value for F is infinitesimally more than PA, and this leads to Equation 2.6 for the maximum work.

Although Equations 2.5 and 2.6 apply to reversible processes, they also apply to some irreversible processes. This will be the case when the actual expansion (or compression), if considered by itself, is quasistatic, but where there is irreversibility elsewhere in the system. This is best illustrated by two examples.

1.   Consider a cylinder equipped with a frictionless piston containing two solids that react slowly to produce a gas, as in Figure 2.5(a). Because the gas is released slowly, the pressure P inside the cylinder is only

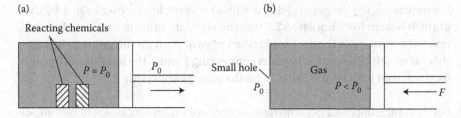

(a)

Reacting chemicals

$P = P_0$ P_0

Small hole   Gas

P_0

(b)

$P < P_0$ F

Figure 2.5   Two examples of $dW = -PdV$ when the process is irreversible. (a) Two solids react to produce a gas. (b) Gas is allowed to leak through a small hole.

infinitesimally greater than the pressure P_0 of the surroundings, and
the piston is always infinitesimally close to mechanical equilibrium.
In other words, there is never a finite pressure drop across the piston.
The chemical reaction is irreversible in that it cannot be reversed by
an infinitesimal change in the external conditions, such as the pressure or the temperature. By the argument presented above, the work
done on the system by its surroundings is again

$$W = -\int_{V_1}^{V_2} PdV = -\int_{V_1}^{V_2} P_0 dV = -P_0\left(V_2 - V_1\right) \qquad 2.7$$

where V_1 and V_2 are the initial and final volumes. This result holds
even though the whole process is irreversible because of the chemical
reaction.

2.   Consider as a second example the pump, shown in Figure 2.5(b), containing gas at the high pressure P. The frictionless piston is pushed in
slowly, thus expelling the enclosed gas through the small hole at the
end into the surrounding atmosphere at the lower pressure P_0. As it
is being pushed in slowly, the piston is always infinitesimally close to
being in mechanical equilibrium, with the applied force being only
infinitesimally greater than PA. Again, there is never a finite pressure
drop across the piston. However, the whole process is irreversible
because, even if F is reduced slightly, the process will not stop, and gas
will still flow out through the small hole. The work done by the compressing piston is again

$$W = -\int_{V_1}^{V_2} PdV$$

where V_1 and V_2 are the initial and final volumes.

We now return to the general discussion of reversible processes. On a PV diagram, it is seen from Equation 2.6 that the absolute value of the work done in a
reversible process is the area under the curve or path for the process. Because
there is an infinite number of paths connecting 1 and 2, the work done depends
on the actual path chosen, that is, on the way P varies with V.

For example, suppose a gas undergoes isothermal expansion, as would be the case
if the cylinder walls were diathermal and were in contact with a thermal reservoir at
T. This path is represented by the upper curve 1–2 shown in Figure 2.6. Then

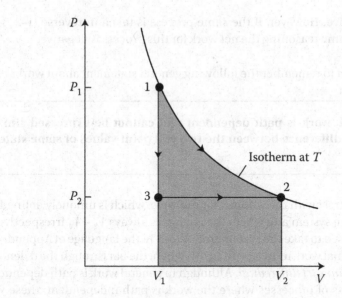

Figure 2.6 Work depends on the path.

$$W = -\int_{V_1}^{V_2} P\,dV = -nRT\int_{V_1}^{V_2}\frac{1}{V}\,dV = -nRT\,\ln\!\left(\frac{V_2}{V_1}\right)$$ 2.8

where we have used the ideal gas law $PV = nRT$. Notice that the result in Equation 2.8 works equally well for expansion and compression. For expansion $V_2 > V_1$ and so $\ln(V_2/V_1) > 0$, and the work done on the gas is negative. This makes sense, because in expansion the gas does positive work on its surroundings. If the gas is compressed from volume V_1 to V_2, then $V_2 < V_1$ and $\ln(V_2/V_1) < 0$, and the work done on the gas is positive, as it must be for compression.

Another simple reversible path is 1–3–2 in Figure 2.6, consisting of an isochoric (constant volume) decrease of pressure 1–3 followed by an isobaric expansion 3–2. For this process, the work done on the gas is simply $-P_2(V_2 - V_1)$ which is different from that for the isothermal expansion.

The relationship between work done and area on a PV diagram is especially useful for a reversible cycle that begins and ends at the same point. For example, consider the process 1–3–2–1 in Figure 2.6. The area enclosed by the path is equal to the magnitude of the net work done on the gas for the entire process. However, the *sign* of the net work depends on the direction taken. Positive work is done on the gas in the process 2–1 and negative work is done in the process 3–2. Because the positive work is greater, the net work for the entire cycle is positive. However, if the same process is taken in reverse (1–2–3–1), then by the same reasoning the net work for this process is negative.

It is important to remember the following general statement about work:

> **In general, work is path dependent and cannot be expressed simply as the difference between the two end point values of some state function.**

This is in contrast with the volume, for example, which is uniquely defined by the state of the system and where the change is always $V_2 - V_1$ irrespective of the process used to take the system from 1 to 2. In the language of Appendix B, the infinitesimal work term is written đW, where the bar through the d denotes that đW is an *inexact differential*. Although in general work is path dependent, there is a class of processes where the work is path independent; these will appear in Chapter 3 in the discussion of adiabatic work.

EXAMPLE 2.2

Consider a process in which a fixed amount of ideal gas is taken along the path 1–2–3–1 shown in Figure 2.6. The isotherm is at $T = 293$ K. Other known parameters are $P_1 = 1.0$ atm, $V_1 = 1.5$ L, and $V_2 = 3.0$ L. Find the work done on the gas in each of the three steps and the net work done on the gas for the cycle.

   Solution: Note that if we use SI units, with $R = 8.315$ J/(mol·K), then $P_1 =$ 1.0 atm = 1.013×10^5 Pa, $V_1 = 0.0015$ m³, and $V_2 = 0.0030$ m³. Immediately we see that with $PV =$ constant along the isotherm requires $P_2 = 0.5$ atm = 5.065×10^4 Pa, and the amount of gas is

$$n = \frac{P_1 V_1}{RT} = \frac{\left(1.013 \times 10^5 \text{ Pa}\right)\left(0.0015 \text{ m}^3\right)}{\left(8.315 \times 10^5 \text{ J}/\left(\text{mol} \cdot \text{K}\right)\right)\left(293 \text{ K}\right)} = 0.06237 \text{ mol}$$

The work done along path 1–2 (the isotherm) is given by Equation 2.8 as

$$W = -nRT = -\left(0.06237 \text{ mol}\right)\left(8.315 \times 10^5 \text{ J}/\left(\text{mol} \cdot \text{K}\right)\right)\left(293 \text{ K}\right)\ln 2 = -105.3 \text{ J}$$

Along path 2–3 the work is

$$W = -P\Delta V = -\left(5.065 \times 10^4 \text{ Pa}\right)\left(-0.0015 \text{ m}^3\right) = +76.0 \text{ J}$$

The work done along 3–1 is zero, so the net work for the cycle is

$$W = -105.3\,J + 76.0\,J = -29.3\,J$$

The negative value is expected, as explained in the text.

2.2.2 Free Expansion

Consider a gas in the state (P, V) contained in the left-hand part of a double-sectioned chamber, as in Figure 2.7. There is a vacuum in the right-hand part. For simplicity, let the volumes of each part be equal to V. Let the intervening partition be broken so that the gas expands to fill the entire space with volume $2V$, quickly settling down to a new equilibrium state. This process is known as a *free expansion*.

Strictly, the walls should also be adiabatic for a true free expansion, but this is not important here.

How much work is done on the gas in this process? A blind application of $W = -\int_{V_1}^{V_2} P dV$ would give a finite answer, because the volume certainly changes. Of course, the answer is zero, since the gas does no work on the surroundings outside the chamber, and there is no outside agent applying any force as with the piston in the previous example. The expression $W = -\int_{V_1}^{V_2} P dV$

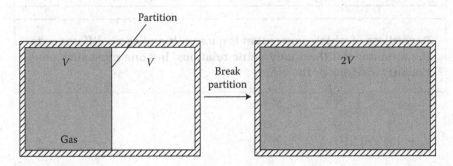

Figure 2.7 A free expansion. When the partition is broken, the gas occupies the whole volume, but no work is performed on the gas.

cannot be applied here, because this process is not reversible. This example illustrates two points in thermodynamics.

1.   It is important to be clear as to what is the system. Here it is the chamber as a whole and not just the left-hand part initially containing all the gas.

2. $dW = -PdV$ is applicable only to reversible processes and to those special irreversible processes such as those considered in Section 2.2.1 where there is no finite pressure drop across the piston.

2.2.3 Sign Convention for Work

There are unfortunately different sign conventions used for W in different texts. The convention we have already established in Section 2.2.1 is that W is the work done on a system by its surroundings. This is the most common convention today and the one that most physicists adopt. However, in some books (especially older ones) you will see the opposite convention, in which W is taken to be the work done *by* a system on its surroundings. This alternate convention might make sense in some applications, for example for engineers concerned about how much work is done by an engine. However, we will stick with our established convention for W, because it makes more sense in most applications and makes it easier to understand the first law of thermodynamics (Chapter 3).

Using the adopted convention, the generalization of Equation 2.5 becomes

$$dW = -PdV\,(\text{reversible}) \qquad\qquad 2.9$$

Regardless of which convention is chosen, it makes no difference to the fundamental thermodynamic relations, but one must stick consistently with one or the other.

2.2.4 Dissipative Work

Suppose there is a viscous fluid that can be stirred, as in Figure 2.8, by the action of the falling weight. Because of the dissipative viscous effects in the fluid, the temperature will rise, and the state of the fluid system will change. This work

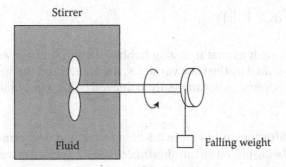

Figure 2.8 Dissipative work.

is performed irreversibly; if the torque on the shaft attached to the stirrer was reduced infinitesimally, the shaft would not start to go the other way with the weight rising. We call this kind of irreversible work *dissipative work*.

Another example of dissipative work is the passing of a current I through a resistor R immersed in the fluid. Then the work performed in time t is $I^2 Rt$. Again this work is irreversible, because reducing the battery voltage slightly will not cause the current to reverse, with the resistor then doing work on the battery.

Unlike reversible work considered in Section 2.2.1, it is not possible to find an expression for dissipative work in terms of the state variables of the system.

2.3 OTHER KINDS OF WORK

There are systems other than the compressible gas or fluid commonly encountered in thermodynamics. In this section you will see the appropriate form for the infinitesimal work term in a reversible process for each of the following.

2.3.1 Extensible Wire

The work done by us acting as the surroundings when a wire at a tension F is stretched through an infinitesimal distance dx is

$$đW = Fdx \qquad\qquad 2.10$$

This is positive for a positive extension dx, which is consistent with the established sign convention.

2.3.2 Surface Film

A surface film, such as that in a soap bubble, has equilibrium states that are completely specified by the state variables, the area A, and the surface tension Γ. Normally Γ is found experimentally to depend on the temperature only and not on the area.

Consider the film shown in Figure 2.9 being stretched isothermally, with the moveable bar being pulled an infinitesimal distance dx by an external force that is only infinitesimally greater than $\Gamma\,\ell$. Then, the infinitesimal work done is

$$dW = \Gamma\ell dx = \Gamma dA \qquad\qquad 2.11$$

The work done is positive if dA is positive. (This analysis assumes a single-sided film.)

2.3.3 Reversible Electrolytic Cell

The equilibrium states of a reversible electrolytic cell, such as the simple Daniell cell, are specified by the state variables, the charge Z stored, and the emf ε. Suppose that Z is increased infinitesimally by dZ; then the work performed by the external charging circuit is

$$dW = \varepsilon dZ \qquad\qquad 2.12$$

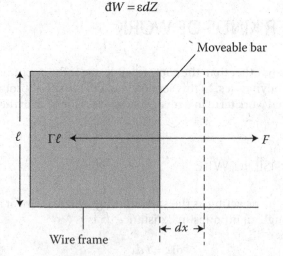

Figure 2.9   The work done is increasing the surface area of a film by dA is ΓdA.

A comment should be added here about notation. It is necessary to use the symbol Z, instead of the more usual symbol Q, for charge because, in thermodynamics, the symbol Q is reserved for heat. Both Z and Q appear in the thermodynamics of an electrolytic cell, and it is important to make a distinction.

2.3.4 A Simple Magnetizable Material

The equilibrium states of a simple magnetic material are specified by the state variables: the overall magnetic moment $\mathcal{M}$, and the applied magnetic field B_0. Paramagnetic and diamagnetic compounds fall into this category, but one must exclude most ferromagnetic materials, where hysteresis effects result in there being no unique relation between $\mathcal{M}$ and B_0 at each temperature. If the magnetization is uniform over the volume of the sample,

$$\mathcal{M} = MV$$

where M is the magnetization, or the magnetic moment per unit volume.

In Appendix C it is shown that, when the sample is uniformly magnetized, the external work required to increase the magnetic moment from $\mathcal{M}$ to $\mathcal{M} + d\mathcal{M}$ in the applied magnetic field B_0 is

$$đW = B_0 d\mathcal{M} \qquad\qquad 2.13$$

2.3.5 A Dielectric Material

The equilibrium states of a dielectric substance are specified by the state variables: the overall electric dipole moment p and the applied electric field E. It is shown in Appendix C, where only linear dielectrics with no hysteresis effects are considered, that the infinitesimal external work required to increase the overall dipole moment of a uniformly polarized dielectric from p to $p + dp$ in a field E is

$$đW = Edp \qquad\qquad 2.14$$

For such a uniformly polarized dielectric, the overall dipole moment is related to the polarization P, or dipole moment per unit volume, by

TABLE 2.2   Infinitesimal Work in Various Reversible Processes

System	Intensive Variable	Extensive Variable	Infinitesimal Work
Gas or fluid	P	V	$-P\,dV$
Film	Γ	A	$\Gamma\,dA$
Cell	ε	Z	$\varepsilon\,dZ$
Magnetic material	B_0	$\mathcal{M}$	$B_0\,d\mathcal{M}$
Dielectric material	E	p	$E\,dp$

$$p = PV$$

The results from all these "other kinds of work" are collected in Table 2.2. Suppose one were to consider as the system only *part* of the original system. Then, if the system was an ideal gas, the pressure of the subsystem considered would be the same as in the original system, but the volume would be smaller. Because the pressure is in this sense size independent, we say that it is an *intensive variable*; conversely, volume is an *extensive variable*. In Table 2.2 the state variables are grouped according to whether they are extensive or intensive.

There is one final point. *All* the work processes that have been considered in this section may be thought of as equivalent to a process whose sole effect on the surroundings is the raising or lowering of a weight, and thus equivalent to mechanical work. Notice that a weight does not actually have to be raised, only that it *could* be raised. Clearly this is so for the case of an expanding gas, as shown in Figure 2.10(a). As another example, consider dissipative electrical work where a current I enters the system containing a resistance R, as in Figure 2.10(b). The current may be thought of as being produced by a generator, the

(a)   (b)

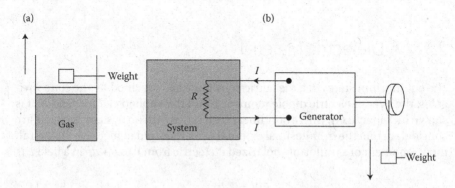

Figure 2.10   Work is always equivalent to raising or lowering a weight. (a) An expanding gas raises a weight. (b) A falling weight produces electric current, which does work at a rate I^2R.

shaft of which is turned by the action of a falling weight. Work is then done on the system at the rate of I^2R.

Similar processes involving the lifting or lowering of a weight may be thought up for all the other forms of work considered. This idea will be useful in distinguishing between work and heat in Chapter 3.

2.4 EXAMPLE OF THE CALCULATION OF WORK IN A REVERSIBLE PROCESS

As an example of a reversible process, consider changing the state of a compressible fluid from (P_1, T_1) to (P_2, T_2). Calculating the work done in this process is a good illustration of some of the ideas developed in this chapter.

The infinitesimal work done in part of the process is $-P \, dV$, so it is necessary to find dV. This example is typical of many in thermodynamics in that one has to find the change in one state function when one is told the change in two others—here $(P_2 - P_1)$ and $(T_2 - T_1)$. A similar example occurred in Section 2.1.4, involving the calculation of ΔF for the wire given the changes in temperature and length. The technique is always the same. For a reversible process, which is a succession of equilibrium states, the equation of state holds at every stage in the process. Thus the equation of state is written in the form that gives the state function whose change we wish to find in terms of the other two whose changes are given.

Therefore,

$$V = V(P, T)$$

By the chain rule:

$$dV = \left(\frac{\partial V}{\partial P} \right)_T dP + \left(\frac{\partial V}{\partial T} \right)_P dT$$

From Sections 2.1.3 and 2.1.2:

$$B = -V \left(\frac{\partial P}{\partial V} \right)_T \quad \text{and} \quad \beta = \frac{1}{V} \left(\frac{\partial V}{\partial T} \right)_P$$

So

$$dV = -\frac{V}{B}dP + \beta V dT$$

then đ$W = -PdV$, or

$$đW = \frac{PV}{B}dP - P\beta V dT$$

and

$$W = \int_{P_1}^{P_2} \frac{PV}{B}dP - \int_{T_1}^{T_2} P\beta V dT$$

In essence, the problem is solved if one can perform the integrations, and this can be done in certain simplified cases. For example, suppose the change is isothermal. Then

$$W = \int_{P_1}^{P_2} \frac{PV}{B}dP = V\left(\frac{P_2^2 - P_1^2}{2B}\right)$$

if the volume V and bulk modulus B stay approximately constant during the process.

Although it is physically reasonable to take V and B outside the integral if their dependencies on P are small, this step can be justified more rigorously mathematically. It is shown in elementary texts on analysis (see the book by Stephenson 1996) that, if $\phi(x)$ and $f(x)$ are two well-behaved functions of x,

$$\int_{x_1}^{x_2} f(x)\phi(x)dx = \phi(\varsigma)\int_{x_1}^{x_2} f(x)dx \qquad \text{where} \qquad x_1 \le \varsigma \le x_2$$

This is known as the second mean-value theorem of integral calculus. If V is a slowly varying function of pressure, then it is justified to take it outside the integral, providing it is given a value between $V(P_1)$ and $V(P_2)$. This will of course differ very little from its original value $V(P_1)$ or simply V. The same argument holds for B.

This is also the rigorous justification for the analysis of the discussion of the stretched wire in Section 2.1.4, where Y, A, and α were taken outside the integral in evaluating ΔF.

Problems

2.1 Ten moles of an ideal gas are compressed isothermally and reversibly from a pressure of 1 atm to 10 atm at 300 K. (a) How much work is done on the gas? (b) How much work is done on the gas in the reverse process?

2.2 Show that for small thermal expansions of a solid object the linear and volume coefficients of expansion are related by $\beta \approx 3\alpha$.

2.3 A concrete road surface consists of 15-m long sections separated by gaps to allow for thermal expansion as the weather changes. Suppose the expected road surface temperatures vary during the year from a low of –15°C to a high of 45°C. (Note: On a sunny day, road surface temperatures can be much higher than air temperatures.) The road is designed so that there remains a 2.0-mm gap between sections on the hottest day. What is the gap on the coldest day?

2.4 Old-style liquid-bulb thermometers used thermal expansion of a liquid inside a tube of fixed diameter to indicate temperature. (a) Consider a mercury-filled thermometer with the liquid expanding along a linear scale, calibrated to a change of 10°C per cm length. If the inside diameter of the tube is 0.10 mm, what is the volume of mercury? Is your answer reasonable? (b) Repeat for a medical thermometer with the same inside diameter, calibrated to change by 1.0°C for every 2.0 cm of length. (c) Repeat parts (a) and (b) using an ethanol-filled thermometer of the same inside diameter and compare your results with the mercury.

2.5 An ideal gas undergoes the following reversible cycle: (i) an isobaric expansion from the state (P_1, V_1) to the state (P_1, V_2); (ii) an isochoric reduction in pressure to the state (P_2, V_2); (iii) an isobaric reduction in volume to the state (P_2, V_1); (iv) an isochoric increase in pressure back to the original state (P_1, V_1). (a) What work is done on the gas in this cycle? (b) If $P_1 = 3.0$ atm, $P_2 = 1.0$ atm, $V_1 = 1.0$ L and $V_2 = 2.0$ L, how much work is done on the gas in traversing the cycle 100 times?

2.6 Two moles of an ideal gas is initially at $P = 1.0$ atm and $T = 300$ K. It is then taken through a three-step reversible process: (i) isobaric expansion to twice its original volume; (ii) isothermal compression, returning to the original volume; (iii) isochoric reduction in pressure to the original state. Find the work done on the gas in each step of the process and the net work done on the gas for the process.

2.7   The bulk modulus of water is 2.2×10^9 N/m^2. (a) Find the pressure needed to compress water by 1% (i.e., to 99% of its initial volume). (b) How much work is done in compressing 10 kg of water by 1%?

2.8   During a reversible adiabatic expansion of an ideal gas, the pressure and volume at any moment are related by $PV^\gamma = c$ where c and γ are constants. Show that the work done by the gas in expanding from a state (P_1, V_1) to a state (P_2, V_2) is

$$W = \frac{P_1V_1 - P_2V_2}{\gamma - 1}$$

2.9   Ice at 0°C and 1 atm has a density of 916.23 kg/m^3, while water under these conditions has a density of 999.84 kg/m^3. How much work is done against the atmosphere when 10 kg of ice melts into water?

2.10 A metal container, of volume V and with diathermal walls, contains n moles of an ideal gas at high pressure. The gas is allowed to leak out slowly from the container through a small valve to the atmosphere at a pressure P_0. The process occurs isothermally at the temperature of the surroundings. Show that the work done by the gas against the surrounding atmosphere is

$$W = P_0\left(nv_0 - V\right)$$

where v_0 is the molar volume of the gas at atmospheric pressure and temperature.

2.11 A hypothetical substance has an isothermal compressibility $\kappa = a/v$ and volume expansion coefficient $\beta = 2bT/v$, where a and b are constants and v is the molar volume. Show that the equation of state is

$$v - bT^2 + aP = \text{constant}$$

2.12 A welded railway line, of length 15 km, is laid without expansion joints in a desert where the night and day temperatures differ by 50°C. The cross-sectional area of the rail is 3.6×10^{-3} m^2. The rails are made of steel, which has $Y = 2.0 \times 10^{11}$ N/m^2. (a) What is the difference in the night and day tension in the rail if it is kept at constant length? (b) If the rail is free to expand, by how much does its length change between night and day?

2.13 Find an expression for the work done when a wire of length L is heated reversibly from a temperature T_1 to a temperature T_2 under conditions of constant tension F.

2.14 The equation of state of a rubber band is

$$F = aT\left[\frac{L}{L_0} - \left(\frac{L_0}{L}\right)^2\right]$$

where L_0 is the original length and a is a constant equal to 1.3×10^{-2} N/K. How much work is performed when the band is stretched isothermally and reversibly from its original length of 10 cm to a length of 20 cm at a temperature of 20°C?

2.15 A block of metal (with $\beta = 5.0 \times 10^{-5}$ K^{-1} and isothermal bulk modulus $B = 1.5 \times 10^{11}$ N/m^2) at a pressure of 1 atm is initially at a temperature of 20°C. It is heated reversibly to 32°C at constant volume. (a) Calculate the final pressure. (b) If the heating had been carried out irreversibly, would this affect your answer?

2.16 A swimming pool measuring 15 m by 10 m by 3.0 m deep is filled with 10°C water. After some time in a hot climate the water reaches 30°C. (a) Assuming a linear temperature variation in the expansion parameter β, how much water (in m^3) has overflowed? (b) In winter the water temperature returns to 10°C. How far will the water level drop?

2.17 0.50 mol of an ideal gas at 1 atm and 293 K is taken along the cycle 1–3–2–1 in Figure 1.6. If $P_2 = 0.80$ atm, find the net work done on the gas for the cycle.

REFERENCE

Stephenson, G., *Mathematical Methods for Science Students*, second edition, Pearson Education, Harlow, 1996.

Chapter 3: The First Law of Thermodynamics

The first law of thermodynamics is essentially a statement of conservation of energy, a familiar and important concept in physics. It gives the precise relationship between the familiar concept of work and the new concepts of internal energy and heat, both of which are defined below. It is useful to begin by approaching the first law historically.

3.1 THE WORK OF THOMPSON AND JOULE

At the beginning of the 19th century, the dominant theory as to the nature of heat was that it was an indestructible substance (*caloric*) that flowed from a hot body, rich in caloric, to a cold body that had less caloric. Heat was quantified by the temperature rise it produced in a unit mass of water, taken as a standard reference substance. The experiments of Joseph Black (professor of medicine at the University of Glasgow, Scotland) in the late 18th century had shown that, when two bodies were put in thermal contact, the heat lost by one in this "method of mixtures" was equal to the heat gained by the other. This suggested that heat was a conserved entity.

A fluid theory similar to the caloric theory was used at the time to explain the flow of electric charge.

However, a different theory was developed by Benjamin Thompson, an American who became Count Rumford of Bavaria and founded the Royal

DOI: 10.1201/9781003299479-3

Institution of Great Britain. Thompson was working in the arsenal in Munich supervising the boring of cannons when he noticed that great heat was produced in that process, as measured by the temperature rise in the cooling water. Further, when he used a blunt boring tool, he found that he could even boil the water, with the supply of heat being apparently inexhaustible. He concluded that heat could not be a finite substance such as caloric, and that there is a direct relation between the work done and the heat produced.

The precise relation between work and heat was established by James Joule some 50 years later in a careful series of experiments between 1840 and 1849. Joule, a Manchester brewer, constructed a tub containing a paddle wheel that could be rotated by the action of weights falling outside the tub, shown schematically in Figure 3.1. Water inside the tub could thus be stirred (irreversibly because of turbulence), raising its temperature between two equilibrium states. The walls of the tub were insulating, so the work was performed under adiabatic conditions. Thus, this kind of work is called *adiabatic work*. Working with extraordinary accuracy, Joule found the following.

1. 4.2 kJ of work is required to raise the temperature of 1 kg of water through 1° Celsius or kelvin (with Joule's British units converted to modern SI units). This is known as the *mechanical equivalent of heat J*. It is interesting to note that, when Joule examined Rumford's results, he obtained a value for *J* that was consistent with his own.

2. No matter how the adiabatic work was performed, it always required the same amount of work to take the water system between the same

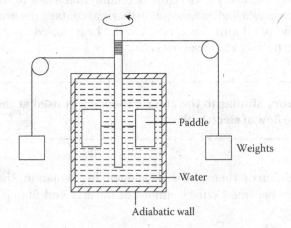

Figure 3.1   A schematic representation of Joule's apparatus.

two equilibrium states. Joule varied his adiabatic work by changing the weights and the number of drops. He also performed the same amount of adiabatic work electrically by allowing the current produced by an electrical generator to be dissipated in a known resistance immersed in the water.

3.2 THE FIRST LAW OF THERMODYNAMICS

The experimental findings are in the following statement of the *first law of thermodynamics*:

> **If a thermally isolated system is brought from one equilibrium state to another, the work necessary to achieve this change is independent of the process used.**

3.2.1 Internal Energy

The statement above is equivalent to saying that the adiabatic work $W_{adiabatic}$ expended in a process is path independent, depending only on the end equilibrium points; this is true whether or not the process is reversible. Therefore there must exist a state function whose difference between the two end points 2 and 1 is equal to the adiabatic work. This state function is called the *internal energy U*, with

$$W_{adiabatic} = U_2 - U_1 \qquad\qquad 3.1$$

In classical mechanics, you are familiar with the idea of the work done on a system increasing the kinetic and potential energies. However, internal energy explicitly excludes any change in these bulk energies. Joule's tub was neither lifted nor set in motion across the floor of the laboratory. From a molecular viewpoint, the external work does in fact go into increasing kinetic and potential energies—those of the individual molecules that have kinetic energy because of their random motion, and potential energy because of their mutual attraction. It is these molecular motions and relative positions that constitute the internal energy U.

3.2.2 Heat

If the system is not thermally isolated, then the work W done in taking the system between a pair of equilibrium points depends on the path. For a given change, $\Delta U = U_2 - U_1$ is uniquely fixed, but W is not now equal to ΔU. In other words, there is a difference between the adiabatic work required to bring about a change between two equilibrium states and the non-adiabatic work required to effect the same change, with the latter having an infinite number of possible values.

The difference between ΔU and W is called the *heat Q*. The generalization of Equation 3.1 is the important mathematical statement of the first law:

First law of thermodynamics:

$$\Delta U = W + Q \qquad\qquad 3.2$$

The first law says that the internal energy can be increased either by doing work on the system or by supplying heat to it. In this form, it is true for all processes whether reversible or irreversible. In Chapter 2 you saw that all forms of work are equivalent to the mechanical raising or lowering of a weight in the surroundings. Therefore:

Heat is the non-mechanical exchange of energy between the system and the surroundings because of their temperature difference.

There has to be a sign convention for heat, just as there is for work. Heat Q is defined to be positive when it enters the system, so Equation 3.2 is correct as it stands, with U increasing if work is done on the system or if heat flows in.

The exchange of energy just described happens quite naturally whenever there is a temperature difference between a system and its surroundings, unless prevented by an adiabatic barrier. This is just what Black observed historically, as described at the beginning of this chapter. The direction of heat flow is *always* from higher to lower temperature. Thus, an alternative definition of heat is the energy that flows spontaneously from an object with higher temperature to an object with lower temperature.

For an infinitesimal process, the first law takes the form

$$dU = đW + đQ$$

where both $đQ$ and $đW$ are written with bars through them to indicate that W, and therefore Q, are, in general, path dependent. In the language of Appendix B, they are inexact differentials. (See also Section 2.2.1.) Although in the special case of adiabatic work $\int đW_{adiabatic}$ is path independent and in that sense $đW_{adiabatic}$ is an exact differential, we shall consistently write the infinitesimal work term as $đW$ with a bar for all cases, because W is not a state function.

> **Although you are accustomed to associating derivatives with changes, resist the notion of thinking of $đQ$ and $đW$ as "changes" in heat or work. The amount of heat and work in a process is always a definite quantity, so it would be meaningless to think of either one as a changing amount. On the other hand, it is reasonable to think of dU as a change in a system's internal energy, resulting from the combined effects of work and heat.**

For a compressible fluid, where $đW = -PdV$ for an infinitesimal reversible process, the first law becomes

$$dU = -PdV + đQ$$

or

$$đQ = dU + PdV \text{ (reversible)} \qquad 3.3$$

It is important to realize that this form applies to a reversible infinitesimal process. That is why reversible appears in parentheses as a reminder.

3.2.3 Heat or Work?

In some cases the distinction between heat and work is obvious. For example, if the gas in Figure 2.4 is contained within adiabatic walls and compressed

or expanded by motion of the piston, any change in internal energy resulting from the piston's motion is solely due to work W. If the process is an isothermal one, the gas must in the same process experience a flow of heat Q
with its surroundings. Under an isothermal compression of an ideal gas, for
which U is a function of T only, W is positive, Q is negative, and the internal
energy is unchanged. Similarly, isothermal expansion of an ideal gas results in
a negative value of W, positive value of Q, and internal energy unchanged. For
non-ideal gases and other compressible fluids, the signs of W and Q are as just
described, but the internal energy does change.

As another example, consider Figure 3.2 showing a gas in a container with
rigid diathermal walls. A current I flows through the heating coils of resistance R wrapped around the container. In Figure 3.2(a), the system denoted
by the dashed line, includes the heating coils. As discussed in Section 2.3.5,
work is being done on the system at the rate I^2R because the current I enters
the system. The energy crossing the system boundary is in the form of work.
In Figure 3.2(b), the system is the gas and container alone, excluding the
coils. Here, no work is done on the system but there is energy flow across the
system boundary in the form of heat because the temperature of the coils is
higher than that of the gas. This simple example shows that in distinguishing
between heat and work, it is very important to be clear as to what constitutes
the system.

A perhaps less obvious example occurs when you use a microwave oven to
raise the temperature of your room-temperature cup of coffee. Does this process involve heat or work? In a case like this, it is good to remember that heat
involves a spontaneous flow of energy from a warmer object to a cooler one.
Since that is not occurring in this example, you must conclude that the microwave oven is doing work.

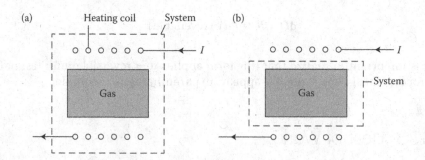

Figure 3.2   An illustration of the difference between work and heat. The input of energy
into the system is in (a) as work and in (b) as heat.

This last example illustrates why some everyday uses of the word "heat" are strictly speaking incorrect in thermodynamics. You might well say "I'm going to heat my coffee," when as the last example shows this is not necessarily the case.

3.3 HEAT CAPACITY

Suppose there is a process that allows heat Q to flow into a system, changing it from one equilibrium state to another with a temperature difference ΔT, as in Figure 3.3. The *heat capacity C* of a system is defined as the limiting ratio of the heat flow divided by the temperature rise:

$$C = \lim_{\Delta T \to 0} \frac{Q}{\Delta T} = \frac{dQ}{dT} \qquad 3.4$$

The *specific heat c* is the heat capacity per unit mass:

$$c = \frac{1}{m}\left(\frac{dQ}{dT}\right) \qquad 3.5$$

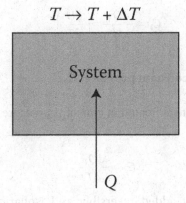

$$T \to T + \Delta T$$

System

Q

Figure 3.3   The heat capacity is $\lim_{\Delta T \to 0}(Q/\Delta T)$.

Another common way to define specific heat is the heat capacity per mole, which is particularly useful in describing gases. This will be presented in Section 3.4.3.

Lower case c is used consistently for specific heat, whether per unit mass, per mole, or another definite quantity.

3.3.1 Constant Volume and Constant Pressure Processes

A process is not completely defined simply by the temperature difference ΔT between the end points. There is a large number of possible reversible paths between these end points, each with a different Q. It follows that there is a large number of possible heat capacities. We shall restrict ourselves to the two that are most commonly used and most useful in thermodynamics.

    1. **Heat capacity at constant volume, C_V.**

Suppose the system is heated under conditions of constant volume, such as a gas in a closed container. For an infinitesimal isochoric reversible process, Equation 3.3 gives đ$Q_V = dU$ where the subscript V on đQ_V indicates that the volume is held constant. Although đQ for this special case is equal to the exact differential dU, the bar is still written through the symbol d because Q is not a state function. Thus

$$C_V = \frac{dQ_V}{dT} = \left(\frac{\partial U}{\partial T}\right)_V \qquad\qquad 3.6$$

    2. **Heat capacity of constant pressure, C_P.**

Now suppose the system is heated at constant pressure. The heat capacity at constant pressure is

$$C_P = \frac{dQ_P}{dT}$$

where đQ_P is the heat added reversibly and isobarically to produce the temperature rise dT. Analogous to the result đ$Q_V = dU$ for constant volume, one can obtain a similar result for $đQ_P$ by defining a new state function, the *enthalpy H*, as

Enthalpy H:

$$H = U + PV \qquad\qquad 3.7$$

For an infinitesimal process

$$dH = dU + PdV + VdP$$

Using the infinitesimal form of the first law, Equation 3.3:

$$dH = đQ + VdP \qquad\qquad 3.8$$

which holds again for a reversible process. For a reversible, isobaric process then,

$$dH = đQ_P \qquad\qquad 3.9$$

That is, the heat flow (in or out) in a reversible isobaric process is equal to the enthalpy change. Thus,

$$C_P = \frac{đQ_P}{dT} = \left(\frac{\partial H}{\partial T}\right)_P \qquad\qquad 3.10$$

Later in this chapter, we present an engineering use for H when considering steady flow processes and dealing with turbines. Enthalpy also has a particular use in chemistry; it will be shown in Chapter 7 that the enthalpy change in an isobaric chemical reaction is equal to the *heat of reaction*, whether or not that reaction is reversible.

How do you decide whether it is more useful to measure C_V or C_P for a particular material? For solids and liquids, it doesn't make much difference. As they are heated, there is normally negligible change in pressure and volume, so there is no significant difference between C_V and C_P, and the specific heat is often expressed as c without a subscript. However, for gases it is important to distinguish between C_V and C_P (or c_V and c_P). This distinction will be explored further in Section 3.5.

3.3.2 Measuring Heat Capacities

Consider again the definition of heat capacity and the requirement that the heat has to be introduced reversibly. This means that, as the heat is introduced, the system must pass through a series of equilibrium states, with the pressure and temperature always being uniform throughout the system. What would be the consequences of putting in a small burst of heat at some point in the system, for example a gas, from a source at a finite temperature above that of the system? This will cause local heating, with pressure and temperature gradients introduced to the gas system, which makes questionable the determination of the heat capacity. In practice, heat capacities are measured in just this way, so there does appear to be an inconsistency. However, if the relaxation time for the attainment of an equilibrium state in the system is much shorter than the timescale of the heating, the system is always so close to an equilibrium state that there are no significant internal pressure and temperature gradients. The irreversible nature of the heating is then of no consequence. In fact, this is the usual situation, and so there is no inconsistency between the definition of heat capacity and its experimental determination.

3.3.3 Experimental Determination

In practice it is common to measure heat capacity (or specific heat) over a small enough temperature range that the quantity being measured is nearly constant. In that case Equation 3.5 shows that heat Q required to raise the temperature of a sample of mass m and specific heat c through a temperature increase ΔT is

$$Q = mc\Delta T \qquad\qquad 3.11$$

For example, it takes 418.6 J of heat to raise the temperature of 0.10 kg of water from 20°C to 21°C. Thus by Equation 3.11 the specific heat of water is

$$c = \frac{Q}{m\Delta T} = \frac{418.6\ \text{J}}{(0.10\ \text{kg})(1.0°\text{C})} = 4186\ \text{J}/(\text{kg}°\text{C})$$

This is a well-known value, and in fact water's reliable specific heat is used as a standard that can be used to determine the specific heats of other materials in the process known as *calorimetry*. For example, suppose you want to determine the specific heat of a piece of metal. Take a known mass of the metal

TABLE 3.1   Specific Heat of Selected Materials at 20°C Unless Indicated

Material	Specific Heat c in J/(kg·°C)
Aluminum	900
Beryllium	1970
Bismuth	123
Brass	380
Copper	385
Gold	126
Iron	449
Lead	128
Mercury	140
Silver	235
Tungsten	134
Zinc	387
Water	4186
Ice (−10°C)	2050
Steel (typical)	500

at one temperature and place it in another known mass of water at a different temperature. Heat flows between the two materials until they reach the same equilibrium temperature. By conservation of energy, the heat lost by the initially warmer material is equal to the heat gained by the cooler one, and Equation 3.11 allows you to determine the unknown specific heat. Of course, this method ignores any heat transferred to or from the environment in the process, so efforts must be made to minimize such transfer. Measured values of specific heat for a few selected materials are given in Table 3.1.

In the old (non-SI) system used in calorimetry, the calorie (cal) was used as a unit of heat (and other forms of energy). It was defined so that at a specific temperature and pressure the specific heat of water would be exactly 1000 cal/(kg·°C), or equivalently exactly 1 cal/(g·°C). Thus, if needed, the conversion between calories and joules is 1 cal = 4.186 J.

EXAMPLE 3.1

A calorimeter contains 100 g of water at 20°C. Then 170 g of a pure dense silvery metal is placed in the calorimeter. The final equilibrium temperature of the metal and water is 26.1°C. Determine the specific heat of the metal.

Solution: In the process of reaching equilibrium, the heat flowing out of the metal is absorbed by the water. Using $Q = mc\Delta T$ for each substance, this means

$$m_m c_m \Delta T_m = m_w c_w \Delta T_w$$

where the subscripts used are m for metal and w for water. This setup assumes we use the absolute value (positive value) for each ΔT. Solving with numerical values,

$$c_m = \frac{m_w c_w \Delta T_w}{m_m \Delta T_m} = \frac{(0.10 \text{ kg})(4186 \text{ J}/(\text{kg} \cdot °\text{C}))(26.1°\text{C} - 20°\text{C})}{(0.17 \text{ kg})(90°\text{C} - 26.1°\text{C})} = 235 \text{ J}/(\text{kg} \cdot °\text{C})$$

If this is a pure metal, as reported, it may be silver, because this value matches the specific heat of silver in Table 3.1. Measuring the metal's density would help to identify it.

3.3.4 Latent Heat

When heat transfer results in a phase transformation, there is no change in temperature (as long as pressure is kept constant). For example, boiling water at 1.0 atm pressure occurs at 100°C and continued heating results in continued boiling with no temperature increase. Therefore, the concept of heat capacity does not apply to such processes. However, the amount of heat applied is directly proportional to the mass transformed from one phase to another, with the constant relating the two called *latent heat L*. That is,

$$L = \frac{Q}{m} \qquad\qquad 3.12$$

with SI units J/kg.

The value of L is different for each phase transformation. For water, the latent heat for melting ice at 0°C is 333.6 kJ/kg, and the latent heat for boiling water at 100°C is 2256 kJ/kg. Evidently, it takes nearly seven times as much energy to boil water than to melt the same amount. This should be consistent with your experience in cooking. On the stovetop you can thaw frozen food relatively quickly, but you can cook noodles or rice until done without boiling away too much of the water.

The phase transformations described above are reversible. To freeze water you have to remove 333.6 kJ/kg. This process is done by heat transfer with the air on a cold day or in a refrigeration system. Similarly, condensing steam requires removal of 2256 kJ/kg. A more familiar process is the evaporation of sweat on your skin, which removes thermal energy from your body. The latent heat for evaporation at body temperature 37°C is about 2420 kJ/kg.

Phase transformations will be discussed in more detail in Chapter 10.

EXAMPLE 3.2

You have a soft drink that can be assumed to have the same thermal properties as water. Your 350-gram drink is at room temperature 20°C, so you cool it by tossing in 100 g of ice at 0°C. Once the mixture reaches equilibrium at 0°C, how much ice remains?

Solution: To go from 20°C to 0°C, the water (soft drink) must give up $Q = m_w c_w \Delta T_w$ with subscript w for water. The same amount of heat melts mass m_I of ice, with $Q = m_I L$. Equating the two expressions for Q gives

$$m_I = \frac{m_w c_w \Delta T_w}{L} = \frac{(0.35\ \text{kg})(4186\ \text{J}/(\text{kg}\cdot\text{°C}))(20\text{°C})}{333{,}000\ \text{J/kg}} = 0.088\ \text{kg}$$

That's 88 g of ice melted, so the remaining amount is a mere 12 g out of the original 100 g.

3.3.5 The Method of Mixtures

At the beginning of this chapter we mentioned that Joseph Black quantified heat using the method of mixtures. Let us see how the first law justifies the idea of "heat lost equals heat gained."

Suppose that two systems, A and B, are put in thermal contact with an adiabatic wall surrounding both systems. For the two systems, the first law is

$$U_f^A - U_i^A = Q^A + W^A$$
$$U_f^B - U_i^B = Q^B + W^B$$

where subscripts i and f refer to the initial and final states and the superscripts A and B refer to the different systems. Adding,

$$\left(U_f^A + U_f^B\right) - \left(U_i^A + U_i^B\right) = \left(Q^A + Q^B\right) + \left(W^A + W^B\right)$$

now $\left(U_f^A + U_f^B\right) - \left(U_i^A + U_i^B\right)$ is the change in the internal energy of the *composite* system, and $(W^A + W^B)$ is the work done on it. Hence $(Q^A + Q^B)$ is the heat that flows into the composite system, which is known to be zero because this composite system is surrounded by an adiabatic wall. So

$$Q^A + Q^B = 0 \qquad \text{and} \qquad Q^A = -Q^B$$

or "heat lost by B equals heat gained by A."

3.4 KINETIC THEORY OF GASES

Some simple but important relations for ideal gases can be obtained by applying classical mechanics to the motion of individual molecules in a gas. This approach is called *kinetic theory* and is valid because it deals with mean values of molecular energy without going into specifics about the statistical distribution of energy, which is covered in Chapter 6.

3.4.1 Mean Energy and Equipartition

Consider again a gas enclosed in the piston-cylinder arrangement in Figure 2.4. You can show by classical mechanics (Problem 3.8) that a molecule of mass m and velocity component v_x (toward the right) exerts a mean pressure $\bar{P}$ on the piston given by $\bar{P}V = mv_x^2$ where V is the volume of gas enclosed. For a gas of N molecules,

$$\bar{P}V = Nm\overline{v_x^2}$$

Using the molecular version of the ideal gas law $PV = Nk_BT$ (where k_B is Boltzmann's constant), the mean kinetic energy associated with the velocity component v_x is

$$\frac{1}{2}m\overline{v_x^2} = \frac{1}{2}k_BT$$

There's nothing distinctive about the x-direction in space, so the results for y and z must be the same, and thus the mean kinetic energy of a molecule of gas is $\bar{K} = 3\left(1/2m\overline{v_x^2}\right)$ or

$$\bar{K} = \frac{3}{2}k_B T \qquad\qquad 3.13$$

Notice that this is just the *translational* energy of the gas molecule. If there are other forms of energy, specifically vibrational and rotational, they must be dealt with separately, as shown in Section 3.4.3.

Equation 3.13 is a particular example of a more general rule, called the *equipartition theorem*:

> **Equipartition theorem: In thermodynamic equilibrium at temperature T, each independent quadratic degree of freedom contributes $1/2\,k_B\,T$ to the mean energy of a molecule.**

In the gas just considered, the x-, y-, and z-components of the velocity constitute the three degrees of freedom, because they are independent of one another and each is quadratic (e.g., v_x^2). In Section 3.4.3 you will see how the equipartition theorem can be applied to other systems.

3.4.2 Molecular Speeds

In an ideal gas, each of the three velocity components is random, and the distributions of velocity components and speeds are statistical problems that will be considered in Chapter 6. However, if one deals only with mean values, Equation 3.13 gives

$$\bar{K} = \frac{3}{2}k_B T = \frac{1}{2}m\overline{v^2} = \frac{1}{2}m\overline{v^2}$$

where in the last step we have used the fact that the factors 1/2 and m are the same for every molecule. This result suggests that a *root-mean-square speed* be defined as $v_{rms} = \sqrt{\overline{v^2}}$, which in this case becomes

$$v_{rms} = \sqrt{\frac{3k_B T}{m}}$$ 3.14

We stress that the root-mean-square speed is merely a typical speed for a gas molecule at temperature T, and that the statistical distribution of speeds is quite significant (see Section 6.4.2). Further, note that in statistics in general $\overline{v^2} \neq (\overline{v})^2$, so the root-mean-square speed is *not* simply the square of the mean speed.

For gases under ordinary conditions, the root-mean-square speed is surprisingly high. For example, for nitrogen gas (N_2, with molecular mass 28.0 u) at room temperature $T = 20°C = 293$ K, the root-mean-square speed is 511 m/s.

EXAMPLE 3.3

Find the rms speed of (a) oxygen and (b) helium in air at room temperature. Compare the two results with each other and with the rms speed of nitrogen at the same temperature (given in the text).

Solution: (a) Oxygen is slightly heavier than nitrogen. With an atomic weight of 16.0 u, the molecular weight of O_2 is 32.0 u. Applying Equation 3.14 with the appropriate constants:

$$v_{rms} = \sqrt{\frac{3k_B T}{m}} = \sqrt{\frac{3(1.381\times10^{-23}\text{ J/K})(293\text{ K})}{(32.0\text{ u})(1.661\times10^{-27}\text{ kg/u})}} = 478\text{ m/s}$$

(b) Helium is a light monatomic gas with a mass of 4.0 u. Its rms speed at the same temperature is

$$v_{rms} = \sqrt{\frac{3k_B T}{m}} = \sqrt{\frac{3(1.381\times10^{-23}\text{ J/K})(293\text{ K})}{(4.0\text{ u})(1.661\times10^{-27}\text{ kg/u})}} = 1352\text{ m/s}$$

Comparing the three gases (oxygen, nitrogen, and helium), the progression of rms speed is as expected, with the heavier molecules generally traveling slower. The square root operator means this progression is not linear. Helium's mass is one-eighth that of the oxygen molecule, but its speed is only $\sqrt{8}$ times greater.

Is this difference in typical speed the reason nitrogen and oxygen are plentiful in Earth's atmosphere, but helium is absent? This is a subtle question. Notice that helium's rms speed is still well below the escape speed from Earth's surface (11.2 km/s) computed by Newtonian dynamics. Thus the question of escape from Earth ultimately depends on the distribution of molecular speeds more than the rms value.

3.4.3 Non-Monatomic Ideal Gases

The most common gases in the atmosphere are the diatomic gases N_2 and O_2, followed by the monatomic gas argon that constitutes less than 1% of the atmosphere. Equation 3.14 gives the root-mean-square speeds for all these gases, because the analysis leading to that result considered only translational motion.

However, if you want to consider all the thermal energy contained in diatomic gases, rotational modes of energy must also be considered. Think of the two atoms in the molecule as connected by a thin rod. The molecule is free to spin about either of the two axes perpendicular to the connecting rod. This adds two rotational degrees of freedom to the three translational degrees for a total of five. By the equipartition theorem, the total thermal energy of a diatomic gas molecule is then $5 \times 1/2\, k_B T = 5/2\, k_B T$.

> The third axis of rotation, along the line connecting the two atoms, is not counted as a degree of freedom in this analysis. This is for a subtle reason. The quantum energy of a rotational state is inversely proportional to the rotational inertia. With most of the atom's mass concentrated in the small atomic nucleus, the rotational inertia along this axis is much smaller than the other two, so the energy needed to excite that rotational state is too large to be realized at room temperature.

The results just quoted are easily verified by measuring the gas's heat capacity. By Equation 3.6, a diatomic gas's heat capacity is $C_V = 5/2\, k_B$ per molecule, which is conveniently expressed as a molar specific heat (i.e., for one mole or N_A molecules) $c_V = 5/2 \, N_A \, k_B = 5/2R$. This has numerical value $c_V = 5/2R = 20.8$ J/(K·mol), in good agreement with measured values. On the other hand, monatomic gases such as argon and helium should have molar heat capacity $c_V = 3/2R = 12.5$ J/(K·mol) according to the equipartition theorem, and experimental measurements verify this.

It becomes problematic to apply the equipartition theorem to more complex gas molecules, such as CO_2 and H_2O. With three atoms in a molecule, the nature of the bonds is important because they undergo some flexing and bending, which can absorb energy and thereby affect specific heat measurements.

3.4.4 Solids

The equipartition theorem also applies well to some solids such as copper, in which the atoms form a cubic lattice. In that case each atom can be treated as a harmonic oscillator in each of three directions. A harmonic oscillator has two degrees of freedom, one for kinetic energy and one for potential energy. According to the virial theorem in classical mechanics, the kinetic and potential energy contribute equal amounts to the net energy. Thus the three directions times two degrees of freedom yield a total of six degrees of freedom, with a predicted molar heat capacity, $c_V = 6 \times 1/2R = 3R$. At low temperatures the measured value is quite close to this. As temperature increases, there is an increasing contribution to the heat capacity from the conduction electrons in copper, reaching about $0.02R$ at room temperature. This is due to quantum mechanical effects, where the conduction electrons must be treated by quantum (Fermi–Dirac) statistics. This effect is considered in Chapter 13.

3.5 IDEAL GASES AND THE FIRST LAW

Another set of simple relations for ideal gases can be obtained from the first law of thermodynamics. An ideal gas is one that obeys the equation of state (the ideal gas law) $PV = nRT$. It is shown in Chapter 8 that, as a consequence of this equation of state, the internal energy is a function of temperature alone. This is consistent with the result from kinetic theory, shown in Section 3.4. By writing

$$U = U(T) \qquad\qquad 3.15$$

three important results for an ideal gas will follow.

3.5.1 Free Expansion

Free expansion was introduced in Section 2.2.2, where it was shown that no external work against the surroundings is performed when a gas expands irreversibly into a larger chamber. Suppose the walls are adiabatic, so no heat enters the system. As $W = 0$ and $Q = 0$, the first law shows that $U_i = U_f$, where the subscripts i and f denote the initial and final equilibrium states. Equation 3.15 implies that $T_i = T_f$, and thus there is no temperature change between the end states. This should make intuitive sense, given the connection between temperature and molecular speeds implied by kinetic theory (Section 3.4). That is, free expansion does not change any of the molecular speeds, so it should not affect temperature.

In 1843, Joule attempted to measure the temperature change in the free expansion of air. He was unable to detect any temperature change, within experimental error. With modern precision equipment, experimenters are able to measure only very small temperature changes; thus air behaves approximately as an ideal gas at normal temperatures. It is found that *all* known gases cool slightly on undergoing a free expansion. This is consistent with the kinetic theory idea that temperature is associated with the kinetic energy of the molecules. If the gas expands, then the intermolecular attraction potential energy goes up as the molecules get further apart. As the total internal energy U is constant for the free expansion, this means that the kinetic energy, and therefore the temperature, goes down.

The quantity $(\partial T/\partial V)_U$ is a measure of the cooling effect occurring in a free expansion and is known as the Joule coefficient μ_J. In Chapter 8 you will see how thermodynamics helps to derive an expression for μ_J from the equation of state, even though free expansion is an irreversible process.

3.5.2 $C_P - C_V$

Consider n moles of an ideal gas. By Equation 3.3, $đQ = dU + PdV$ for an infinitesimal reversible process. The constant-volume heat capacity is $C_V = (\partial U/\partial T)_V$ and, for the special case of an ideal gas where $U = U(T)$, this becomes

$$C_V = \frac{dU}{dT} \quad \text{or} \quad dU = C_V dT \qquad \text{(ideal gas)}$$

In other words, in the special case in which $U = U(T)$, the partial derivative in $C_V = (\partial U/\partial T)_V$ may be replaced by the ordinary derivative in $C_V = dU/dT$, or $dU = C_V dT$.

Using this result, the infinitesimal form of the first law becomes

$$Q = C_V dT + P dV$$

Now consider a constant-pressure process. Dividing all through by dT and taking the partial derivative at constant P

$$\frac{dQ_P}{dT} = C_P = C_V + P\left(\frac{\partial V}{\partial T}\right)_P$$

Differentiating $PV = nRT$ with respect to temperature, $P(\partial V/\partial T)_P = nR$, and it follows that

$$C_P = C_V + nR \qquad \text{(ideal gas)} \qquad\qquad 3.16$$

This relation between C_P and C_V is very simple for an ideal gas; another result for $C_P - C_V$ is derived for a general system in Chapter 8.

Equation 3.16 is confirmed by experimental measurements. Recall from Section 3.4.3 that in theory specific heat $c_V = 3/2R$ for monatomic gases and $c_V = 5/2R$ for diatomic gases. According to Equation 3.16, the molar specific heats c_V and c_P are related by $c_P = c_V + R$. Therefore, theory predicts $c_P = 5/2R$ for monatomic gases and $c_P = 7/2R$ for diatomic gases. Measured values are very close to these predictions.

3.5.3 The Equation of an Adiabat

Throughout an isothermal reversible expansion at temperature T, the pressure and volume of an ideal gas are always related by $PV = nRT$. There is a very simple relation between P and V if the expansion is performed both adiabatically and reversibly. From Equation 3.3

$$Q = dU + PdV \qquad \text{(reversible)}$$

so

$$0 = dU + PdV \qquad \text{(reversible and adiabatic)}$$

or

$$0 = C_V dT + PdV$$

because $dU = C_V\, dT$.

But the equation of state $PV = nRT$ holds at all points in this expansion, although T is no longer constant as it was for isothermal expansion. Therefore one may substitute $P = nRT/V$ to obtain

$$0 = C_V dT + \frac{nRT}{V} dV$$

or

$$0 = \frac{C_V}{T} dT + \frac{nR}{V} dV$$

Integrating

$$C_V \ln T + nR \ln V = \text{a constant*}$$

Dividing through by C_V and using Equation 3.16,

$$\ln T + (\gamma - 1) \ln V = \text{a constant*}$$

where the ratio of the heat capacities C_p / C_V is written as γ, called the *adiabatic exponent*. Hence

$$TV^{\gamma - 1} = \text{a constant*}$$

Using the equation of state again, the result is

---

* The constants here are all different.

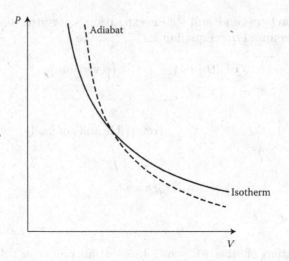

Figure 3.4   The adiabat for an ideal gas has a slope γ times that for the isotherm at each point on the *PV* diagram.

$$PV^{\gamma} = \text{a constant*} \qquad\qquad 3.17$$

which is the equation for a *reversible adiabat*.

It is a simple matter to show by differentiation that the slope of the adiabat at a particular point (P, V) on the *PV* diagram is γ times that of the isotherm through that point. The result of the previous section shows that $\gamma > 1$, because nR is a positive number, and so the adiabat has the steeper slope. This is shown in Figure 3.4.

An *adiabat* (noun) is the curve that describes an *adiabatic* (adjective) process.

3.5.4 Non-Ideal Gases

In a real gas, there are intermolecular attractions. Additionally, the molecules themselves occupy a finite volume. The equation of state therefore has to be modified from the simple $PV = nRT$ for an ideal gas. The most successful modification is that of van der Waals. His equation is

$$\left(P + a/\upsilon^2\right)\left(V - nb\right) = nRT \qquad\qquad 3.18$$

where the first factor on the left contains the modification a/υ^2 to the pressure due to the molecular interactions, and the second factor contains the modification nb to the volume to take into account the molecular volume (a and b are constants). The reader is referred to other texts for a discussion of the a/υ^2 term. Notice that the first factor contains the molar volume $\upsilon = V/n$ and not the total volume V; otherwise, the left-hand side would not be extensive.

3.6 THE JOULE–KELVIN EFFECT

Although general flow processes are discussed in the next section, and again in Chapter 8, it is instructive to formulate some of the basic ideas at this stage. The Joule–Kelvin effect is used in the liquefaction of gases and is often called the *throttling process*, for reasons that will be immediately obvious.

The process is illustrated in Figure 3.5(a). Gas is forced at a constant pressure and at a steady rate through a small hole, or series of holes, to emerge at a constant pressure. The series of small holes is usually in the form of a plug of cotton, wool, or similar material. There is a finite pressure drop across the plug, making the process irreversible. The walls of the chamber are thermally insulating, and so the process is also adiabatic.

In order to analyze this process, consider a given mass of gas as it passes through the plug. Imagine this gas as shown in Figure 3.5(b) being initially contained in a cylinder in the equilibrium state (P_i, V_i) and slowly being forced at constant pressure P_i though the plug. As the gas emerges from the plug at the pressure P_f, imagine it pushing back the piston of another cylinder until all the gas has passed through the plug. This creates the situation as in Figure 3.5(c), with the gas finally in the equilibrium state (P_f, V_f). It is important to realize that this device with the pistons is only to help us with our analysis; the real process is not a "one-shot" one as here, but rather is a steady flow process as in Figure 3.5(a).

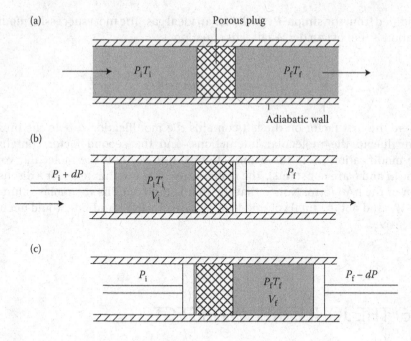

Figure 3.5   A schematic representation of the throttling process.

Because there is no finite pressure drop across the left-hand piston, and as there is no friction, the work done on the gas in forcing the gas through the plug is

$$W = -\int_{V_i}^{0} P_i dV = P_i V_i$$

Although the whole process is irreversible, it has been possible to use the $-PdV$ expression for the work done because the argument used in the original derivation in Chapter 2 is valid here—namely that the external force necessary to push in the piston is still PA. The only finite pressure drop is across the plug and not across the piston. Similarly, the work done by the gas on expanding into the right-hand cylinder is $P_f V_f$.

Applying the first law to the gas,

$$U_f - U_i = 0 + P_i V_i - P_f V_f$$

because no heat enters. So

$$U_i + P_i V_i = U_f + P_f V_f$$

or

$$H_i = H_f \qquad \text{(Iselthalpic process)} \qquad 3.19$$

The throttling process is thus an *isenthalpic* one.

The quantity $(\partial T/\partial P)_H$ is a measure of the temperature change occurring in a throttling process and is known as the *Joule-Kelvin coefficient* μ_{JK}. In Chapter 8 an argument is given using use thermodynamics to calculate μ_{JK} even though the process is irreversible. Unlike the Joule expansion, where there is always cooling, both heating and cooling can occur in the throttling process.

3.7 STEADY FLOW PROCESS—THE TURBINE

This chapter concludes with a brief discussion of steady flow processes, which are of particular importance in engineering. A steady flow device is the flow of a fluid at a constant rate through a device so that some of the internal energy of the fluid is transformed into mechanical work. Some examples of where this occurs are inside an air compressor, a refrigerator, or a turbine. Figure 3.6 shows a general steady flow process.

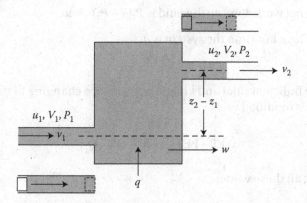

Figure 3.6   A schematic representation of a general steady flow process.

TABLE 3.2   Relevant Parameters in a Steady Flow Device

	Entering Device	Leaving Device
Pressure	P_1	P_2
Specific volume	V_1	V_2
Height	z_1	z_2
Flow velocity	v_1	v_2
Specific internal energy	u_1	u_2

As in the throttling process, it is useful to focus attention on a unit mass of the fluid flowing through the device. The unit mass is considered to be the system. The relevant parameters, which are specific values for the extensive quantities in that they refer to unit mass, are listed in Table 3.2. (In this case V is used rather than v for specific volume, so it is not confused with the flow velocity v.)

In order to determine the specific work (per unit mass) done by the fluid, imagine the unit mass of fluid being contained in a cylinder and being forced at constant pressure P_1 into the device, just as in the throttling process. The work done on the fluid is then P_1V_1, as before. Similarly the work done by the fluid on emerging from the device is P_2V_2. Let the device perform, in addition, the work w (e.g., the shaft work done by a turbine), and heat q enters the system. The following list summarizes all the energy changes, the work performed, and the heat flow:

1.   The internal energy changes by $u_2 - u_1$.

2.   The bulk kinetic energy changes by $\frac{1}{2}\left(v_2^2 - v_1^2\right)$.

3.   The bulk potential energy changes by $g(z_2 - z_1)$.

4.   The net work done *on* the fluid is $P_1V_1 - P_2V_2 - w$.

5.   The heat flow into the system is q.

Because the bulk potential and kinetic energies are changing in this process, the first law is modified to

$$\Delta\left(\text{KE} + \text{PE}\right)_{\text{bulk}} + \Delta U = W + Q$$

Substituting all these values,

$$\frac{1}{2}v_2^2 - \frac{1}{2}v_1^2 + g\left(z_2 - z_1\right) + u_2 - u_1 = P_1V_1 - P_2V_2 - w + q$$

Remembering that the specific enthalpy $h = u + PV$, the shaft work is

$$w = h_1 - h_2 + \frac{1}{2}\left(v_1^2 - v_2^2\right) + g\left(z_1 - z_2\right) + q \qquad 3.20$$

This is the general energy equation for steady flow.

The values for h at different temperatures and pressure are tabulated for different substances in engineering "heat tables." Equation 3.20 can be used to compute w for different flow systems. We shall consider just two important constant flow processes.

3.7.1 The Turbine

Although the temperature of a gas turbine is considerably higher than that of the surroundings, the gas flow is so rapid that only a small quantity of heat is lost by each unit mass of gas, so it is reasonable to approximate $q = 0$. Also, there is usually no difference in elevation at each end. Hence, Equation 3.20 becomes

$$w = h_1 - h_2 + \frac{1}{2}\left(v_1^2 - v_2^2\right) \qquad 3.21$$

This simple result says that the work obtainable can be calculated from knowledge of the enthalpy difference and velocities of the gas entering and leaving the turbine.

3.7.2 Flow through a Nozzle

When a gas flowing through a pipe encounters a change in the cross-sectional area, there is a change of gas velocity. This effect is used frequently in engineering and in particular in a turbine where the gas is "thrown" onto the turbine blades with a high velocity. The incoming gas (steam in the case of a steam turbine) is speeded up by passing it through a nozzle, as in Figure 3.7. No shaft work w is done, the system is assumed to be horizontal, and it is assumed that no heat q enters the system, as the gas flow is too rapid for this to be appreciable. Equation 3.20 then becomes:

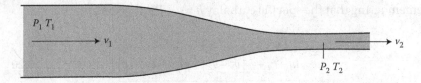

Figure 3.7   A steady flow process through a nozzle.

$$v_1^2 - v_2^2 = 2(h_2 - h_1) \qquad 3.22$$

which relates the velocity change to the enthalpy change.

In practice, it is customary to know the "upstream" conditions P_1, T_1 which means that h_1 is known. However, if as is usual, only the "downstream" pressure P_2 is specified, there is insufficient information to determine h_2. If the flow through the nozzle is assumed to be reversible as well as adiabatic and the gas is treated as ideal, then the downstream temperature T_2 can be found from the adiabatic relation

$$\left(\frac{T_1}{T_2}\right)^\gamma = \left(\frac{P_1}{P_2}\right)^{\gamma-1}$$

This gives sufficient information to find h_2 and therefore v_2.

Problems

3.1   Liquid is stirred at constant volume inside a container with adiabatic walls. The liquid and the container are regarded as the system. (a) Is heat being transferred to the system? (b) Is work being done on the system? (c) What is the sign of the internal energy change of the system?

3.2   Water inside a rigid cylindrical insulated tank is set into rotation and left to come to rest under the action of viscous forces. Regard the tank and the water as the system. (a) Is any work done by the system as the water comes to rest? (b) Is there any heat flow to or from the system? (c) Is there any change in the internal energy of the system?

3.3   A combustion experiment is performed on a mixture of fuel and oxygen contained in a constant-volume container surrounded by a water bath. The temperature of the water is observed to rise. Regard the matter inside the container as the system. (a) Has work been done on

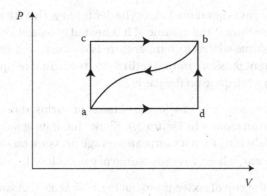

Figure 3.8 *PV* diagram for the processes described in Problem 3.4.

the system? (b) Has heat been transferred between the system and the surroundings? (c) What is the sign of the internal energy change of the system?

3.4   A gas is contained in a cylinder fitted with a frictionless piston and is taken from state a to state b along the path acb shown in Figure 3.8. 80 J of heat flows into the system, and the system does 30 J of work. (a) If instead the work done by the gas system is only 10 J along adb, how much heat flows into the system? (b) When the system is returned from b to a along the curved path, the work done on the system is 20 J. What is the heat transfer? (c) If $U_a = 0$ and $U_d = 40$ J, find the heat absorbed in the processes ad and db.

3.5   An electrical resistance coil, wired to the surroundings, is placed inside a cylinder fitted with a frictionless piston and containing an ideal gas. The walls of the cylinder and the piston are adiabatic. A current of 5.0 A is maintained through the resistance, across which there is a voltage drop of 100 V. The piston is opposed by a constant external force of 5000 N. (a) At what speed must the piston move outward in order that there is no change in the temperature of the gas? (b) Is the electrical energy transferred to the gas as heat or work? (c) Suppose now that the walls are diathermal and the resistance coil is wrapped around the outside of the cylinder. Regard the system as the cylinder and the gas, excluding the heating coil. Is the energy transfer now heat or work?

3.6   Two moles of a monatomic ideal gas are at a temperature of 300 K. The gas expands reversibly and isothermally to twice its original volume. Calculate the work done on the gas, the heat supplied, and the change in the internal energy.

3.7   An ideal gas is contained in a cylinder fitted with a frictionless piston at the pressure P and volume V. It is heated quasistatically and at constant volume so that its temperature is doubled, and then it is cooled at constant pressure until it returns to its original temperature. Show that the work done on the gas is PV.

3.8   Consider a single gas molecule of mass m enclosed in the piston-cylinder arrangement in Figure 2.4. Show that if its velocity component to the right is v_x, then it exerts an average pressure on the piston given by $\bar{P}V = mv_x^2$ where V is the volume of gas enclosed.

3.9   Your 200-g cup of coffee is too hot to drink at 95°C. Assume coffee has the same thermal properties as water. (a) Suppose you want to cool the coffee by adding 20°C water. How much water do you need to add so that the mixture reaches equilibrium at 75°C? (b) Instead of water you use ice initially at 0°C to cool the coffee. How much ice is needed to bring the mixture to equilibrium at 75°C?

3.10   A runner uses energy at an average rate of 15 kcal/min. Assume all the energy is dissipated through the cooling process of evaporating sweat on the skin. How much water does the runner need to consume during the course of a 3.0-hour marathon race? Assess whether your result is reasonable.

3.11   On a hot day you want to cool 0.5 L of tap water initially at 25°C by adding ice. (a) If you add 45 g of ice, what is the final equilibrium temperature of the mixture? (b) How much ice would you need to add to the original sample of water (0.5 L at 25°C) so that the last of the ice melts as the mixture reaches 0°C?

3.12   A 2.5-kg piece of hot lead at 95°C is dropped from a 35-m tower into 10 L of water at 20°C. Compute separately the temperature rise in the water (once equilibrium is reached) due to the lead's (a) thermal properties and (b) gravitational energy. (c) Compare and assess the results in (a) and (b).

3.13   You have a pure metal that is unknown except for the fact that it happens to be among those listed in Table 3.1. A 36.7-g piece of this metal at 50.0°C is placed in a calorimeter containing 150 g of water, initially at 10.0°C. The final equilibrium temperature in the calorimeter is 12.0°C. What is the metal?

3.14   2.5 moles of a monatomic ideal gas is initially at $T = 300$ K and $P = 1.0$ atm. The gas is then taken on a three-step cycle: (i) The pressure and volume increase in such a way that P is proportional to V, until $P = 2.0$ atm; (ii) pressure is reduced at constant volume to 1.0 atm; and (iii) volume is reduced at constant pressure until the initial state is reached. (a) Find the internal energy and volume occupied by the gas in its initial state. (b) Find ΔU, W, and Q for each step in the process. (c) Find the net values of ΔU, W, and Q for the entire cycle. (d) Explain why the signs on your answers in (c) make sense.

3.15 Find the change in the internal energy of one mole of a monatomic ideal gas in an isobaric expansion at 1 atm from a volume of 5 m³ to a volume of 10 m³. (Note that γ for a monatomic ideal gas is 5/3.)

3.16 The molar specific heat of many materials at low temperatures is found to obey the Debye law $c_V = A[T/\theta]^3$ where A is a constant equal to 1.94×10^3 J/(K·mol) and with the Debye temperature θ taking different values for different materials. For diamond it is 1860 K. (a) Evaluate c_V at 20 K and 100 K. (b) How much heat is required to raise one mole of diamond from 20 K to 100 K? (c) What is the average molar specific heat in this range?

3.17 Show that the adiabat curve for an ideal gas is steeper by a factor of γ than the isotherm through a point on the PV diagram.

3.18 Show that the following relations hold for a reversible adiabatic expansion of an ideal gas:

$$TV^{\gamma-1} = a\,\text{constant}$$

$$\frac{T}{P^{1-1/\gamma}} = \text{another constant}$$

The fireball of a uranium fission bomb consists of a sphere of gas of radius 15 m and temperature 300,000 K shortly after detonation. Assuming that the expansion is adiabatic, and that the fireball remains spherical, estimate the radius of the ball when the temperature is 3000 K. (Take $\gamma = 1.4$ for air.)

3.19 A gas with adiabatic exponent γ is compressed adiabatically from an initial state (P_i, V_i) to final state (P_f, V_f). (a) Show that the work done in this process is

$$W = \frac{P_i V_i}{\gamma - 1}\left[\left(\frac{V_i}{V_f}\right)^{\gamma-1} - 1\right]$$

(b) Evaluate the result numerically for one mole of helium gas initially at $P = 1.0$ atm and $T = 300$ K compressed to half its initial volume. (c) Compute the work done in an isothermal compression from the same initial point to half the initial volume. Explain the difference between the numerical results for work done in adiabatic and isothermal compression.

3.20 An interstellar cloud, made up of an ideal gas, collapses with its radius decreasing as

$$R = 10^{13} \left(\frac{-t}{216} \right)^{2/3} \text{ m}$$

with t measured in years. The time t is taken to be zero at zero radius, so that t is always negative. The cloud collapses isothermally at 10 K until its radius reaches 10^{13} m. It then becomes opaque so that from then on, the collapse takes place adiabatically ($\gamma = 5/3$) and reversibly. How many years does it take for the temperature to rise by 800 K measured from the time the cloud reaches a radius of 10^{13} m?

3.21 A thick-walled insulating chamber contains n_1 moles of helium gas at a high pressure P_1 and temperature T_1. The gas is allowed to leak out slowly to the atmosphere at a pressure P_0 through a small valve. Show that the final temperature of the n_2 moles of helium left in the chamber is

$$T_2 = T_1 \left(\frac{P_0}{P_1} \right)^{1-(1/\gamma)} \quad \text{with } n_2 = n_1 \left(\frac{P_0}{P_1} \right)^{1/\gamma}$$

(Hint: Consider the gas that is ultimately left in the chamber as your system.)

3.22 Calculate the work done by a van der Waals gas, with Equation 3.18 as the equation of state, in expanding from a volume V_1 to a volume V_2: (a) at constant pressure P; (b) at constant temperature T.

3.23 A magnetic salt obeys the Curie law

$$\frac{\mu_0 M}{B_0} = \frac{C}{T}$$

where M is the magnetization, B_0 is the applied magnetic field in the absence of the specimen, C is a constant, and μ_0 is the permeability of free space. The salt is magnetized isothermally from a magnetization M_1 to M_2. You may assume that the magnetization is uniform over the volume V of the salt. Show that the work of magnetization is

$$W = \frac{V\mu_0 T}{2C}\left(M_2^2 - M_1^2\right)$$

3.24 The infinitesimal work done in charging a cell is $dW = \varepsilon dZ$ so the rate of doing work is $\varepsilon dZ/dt = \varepsilon I$ where I is the current supplied. A battery is charged by applying a current of 40 A at 12 V for 30 minutes. During this charging process the battery loses 200 kJ of heat to the surroundings. By how much does the internal energy of the cell change, assuming that there are no forms of work other than electrical work?

3.25 A steam turbine takes in steam at the rate of 6000 kg/h and its power output is 800 kW. Neglect any heat loss from the turbine. Find the change in the specific enthalpy of the steam as it passes through the turbine if (a) the entrance and exit are at the same elevation and the entrance and exit velocities are negligible or (b) the entrance velocity is 50 m/s and the exit velocity is 200 m/s, with the outlet pipe 2.0 m above the inlet.

3.26 (a) At what rate do you need to supply heat to melt 400 g of ice at 0°C in five minutes? (b) Now you continue to supply heat at the same rate. How much time does it take to raise the temperature of the melted water from 0°C to 100°C? How much more time does it take to turn all the water to steam at 100°C?

BIBLIOGRAPHY

Adkins, C.J., *An Introduction to Thermal Physics*, Cambridge University Press, Cambridge, 1987.

Reynolds, O., *Biography of James Prescott Joule (History of Physics)*, Wexford College Press, Palm Springs, CA, 2007.

Chapter 4: The Second Law of Thermodynamics

The first law of thermodynamics says that in any process, energy is conserved. Energy may be converted from one form to another, but the total amount of energy is unchanged. The second law of thermodynamics imposes limits on the efficiency of processes that convert heat into work, such as in a steam engine or internal combustion engine. It allows for the definition of a thermodynamic temperature scale, which is independent of the nature of the thermometric substance. The second law also leads to the concept of entropy, which is related both to bulk processes and to the microscopic arrangements within a system.

Before considering the second law, it is useful to discuss Carnot cycles, which are central to the discussion. Then more practical cycles can be considered.

4.1 CARNOT CYCLES

At the beginning of the 19th century, when steam engines were relatively new, there was enormous interest in how their efficiency could be increased. An intellectual giant in this field was the French engineer Sadi Carnot, who in 1824 published an influential paper on how work could be produced from sources of heat. He knew that work could be obtained from an engine if there were heat sources at different temperatures—the boiler and the surrounding air in the case of a steam engine. He also knew that it was possible for heat to flow from a hot body to a cold body with no work being performed, the flow continuing until thermal equilibrium was attained. Carnot realized then, since any return to thermal equilibrium could be used to produce work, any return to equilibrium without the production of this work must be considered a loss. Thus, any temperature difference between two interacting entities may be utilized in the production of work, or it may be dissipated in a spontaneous flow of heat. Carnot concluded that, in an efficient engine, all heat flow between the hot or cold body and the working substance should occur when

DOI: 10.1201/9781003299479-4

these two entities have nearly equal temperatures. This suggested that the working substance follows an isotherm for the times when heat flows and follows an adiabat at other times.

With these ideas in mind, he designed an idealized engine of fundamental significance. The cycle for the *Carnot engine* is depicted in Figure 4.1. Heat flows through a working substance, which for convenience can be assumed to be an ideal gas. The gas is taken through the four-step reversible cycle, abcda, as shown in Figure 4.1. The first step, ab, is an isotherm at temperature T_1, and to maintain constant temperature heat Q_1 enters from a heat reservoir at T_1. Step cd is an isotherm at a lower temperature T_2, where heat Q_2 is rejected to another reservoir at that temperature. The second and fourth steps (bc and da) are adiabats. As shown in Section 2.2.1, the net work W done by the gas in the cycle is the area enclosed by abcda.

It is important to emphasize that a Carnot engine operates between only two reservoirs and that it is reversible. Also, if a working substance other than an ideal gas is chosen, then the shape of the Carnot cycle is different than the one depicted in Figure 4.1, because the equations for the adiabats and isotherms are no longer $PV^\gamma = $ constant and $PV = nRT$, respectively.

> **It is interesting to note that Carnot's ideas were conceived using the caloric concept of heat, before the first law was formulated.**

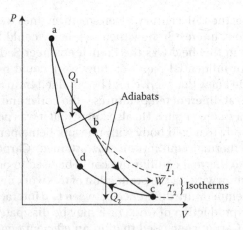

Figure 4.1   Carnot cycle for an ideal gas. The directions of the arrows denote a Carnot engine.

4.2 EFFICIENCY OF AN ENGINE

The Carnot engine described in Section 4.1 is a specific example of a more general class of engines called *heat engines*. In general:

> **A heat engine is an engine that uses the flow of heat from a higher-temperature reservoir T_1 to a lower-temperature reservoir T_2 to do work.**

Any heat engine E may be represented by the schematic diagram in Figure 4.2, where the heat *supplied* Q_1 and the heat *rejected* Q_2 are not necessarily obtained from just two heat reservoirs as in the special case of the Carnot engine. W is the work done by the engine. The arrow around the edge of the block depicting the engine indicates that the engine works in a cycle.

To quantify the notion of engine efficiency, consider it to be a ratio of "what you get out to what you put in." For the engine cycle depicted in Figure 4.2, the output is the work W and the input is the heat Q_1 entering from the hot reservoir. Therefore, the efficiency (the Greek letter η) is

$$\eta = \frac{W}{Q_1}$$

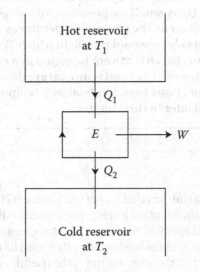

Figure 4.2   Schematic representation of a heat engine working in a cycle. The efficiency is $\eta = W/Q_1 = 1 - Q_2/Q_1$.

Note that efficiency η is a dimensionless quantity that must be less than 1, because the heat Q_2 expelled to the cold reservoir is not available to do work, and therefore $W < Q_1$.

Applying the first law to the working substance in the engine,

$$\Delta U = Q_1 - Q_2 - W$$

This statement of the first law respects the sign convention where heat *into* the system and the work done *on* the system are both counted positively. For an entire cycle the working substance is unchanged. Therefore, $\Delta U = 0$, and the first law becomes

$$W = Q_1 - Q_2$$

Therefore, the engine's efficiency is $\eta = W/Q_1 = (Q_1 - Q_2)/Q_1$ or

$$\eta = 1 - \left(\frac{Q_2}{Q_1} \right) \qquad \qquad 4.1$$

It appears that the strategy for making a more efficient engine is to make the ratio Q_2/Q_1 as small as possible. You might imagine that Q_2/Q_1 depends somehow on the ratio of temperatures T_2/T_1, and therefore a more efficient engine would be one in which T_2/T_1 is also small. That is generally true, but this cannot be stated as a rule, because the exact connection between heat and temperature depends on the type of engine. The connections between heat and temperature in a heat engine are explored later in this chapter.

EXAMPLE 4.1

A nuclear power plant with an efficiency of 0.31 produces electrical energy at a rate of 625 MW. Assume that the nuclear plant operates like a heat engine. For each day the plant operates, find the amount of energy generated by the nuclear fuel and the amount of waste energy that must be expelled.

Solution: The electrical energy output is the useful work generated, so it is equivalent to the quantity W. One day = 86,400 s, so

$$W = 625 \times 10^6 \text{ W} \times \frac{86400 \text{ s}}{\text{day}} = 5.4 \times 10^{13} \text{ J/day}$$

The energy generated by the fuel is the quantity designated Q_1 in the text, the energy input. Thus in one day

$$Q_1 = \frac{W}{\eta} = \frac{5.4 \times 10^{13} \text{ J}}{0.31} = 1.74 \times 10^{14} \text{ J}$$

The waste energy is the difference between the input and the work done:

$$Q_2 = Q_1 - W = 1.74 \times 10^{14} \text{ J} - 5.4 \times 10^{13} \text{ J} = 1.20 \times 10^{14} \text{ J}$$

This is an enormous amount of energy and presents a practical problem for nuclear engineers. Expelling this much energy into a lake or river can cause an unnatural and hazardous increase in water temperature. This is why you often see specially designed cooling towers adjacent to a nuclear plant. The towers are designed to dissipate waste energy more safely.

4.3 STATEMENTS OF THE SECOND LAW OF THERMODYNAMICS

There are two statements of the second law of thermodynamics that are both based on general experience and observations about how real engines work. They were each formulated in the 1850s by Clausius and Kelvin, but the latter was subsequently modified by Planck. (Rudolf Clausius was a German physicist. The person generally referred to as "Kelvin" was William Thomson, honored late in life with the title Lord Kelvin. Max Planck was a German physicist who worked in the late 19th and well into the 20th century.) After stating both forms of the second law, we show that the two statements are equivalent.

4.3.1 Kelvin–Planck Statement

Kelvin–Planck statement of the second law: It is impossible to construct a device that, operating in a cycle, will produce no effect other than the extraction of heat from a single body at a uniform temperature and produce an equivalent amount of work.

From this point we adopt the common practice of referring to the Kelvin–Planck statement simply as the Kelvin statement.

Schematically this statement is represented in Figure 4.3(a). The second law implies that some heat must also be rejected by the device to a body at a lower temperature, as in the heat engine in Figure 4.2. Otherwise, as can be seen from Equation 4.1, one could have an engine with 100% efficiency. If the Kelvin statement were untrue, a number of (impossible) consequences would result. For example, you could drive a ship across the sea just by extracting heat from the sea and converting it entirely into work. Electric lights could be lit using thermal energy from the surrounding air. In short, one could construct a so-called *perpetual-motion machine* (or *perpetuum mobile*). But this cannot be done, because the Kelvin statement seems to be valid in general.

There are some key words and phrases in this statement that need further discussion.

a *Cycle* requires that the state of the working substance is the same at the start and end of the process, although it may change anywhere between these end points. In other words, there is no <u>net</u> change in the state of the working system. Many processes can be thought of that convert heat completely into work, but in all of them there is a net change in the state of the working system. For example, one could heat one mole of an ideal gas and allow it to expand quasistatically and isothermally (by keeping

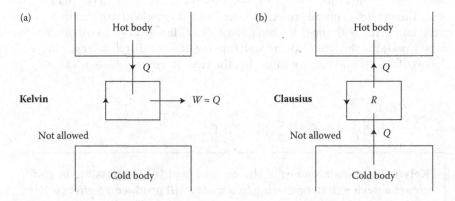

Figure 4.3 (a) Schematic representation of the Kelvin–Planck statement of the second law. (b) Schematic representation of the Clausius statement.

it in contact with a thermal reservoir) from a volume V_1 to $V_2 > V_1$ as in Figure 4.4. The work done by the gas is

$$W = \int_{V_1}^{V_2} PdV = RT \int_{V_1}^{V_2} \frac{1}{V} dV = RT \ln\left(\frac{V_2}{V_1}\right) > 0$$

Because the expansion is isothermal, $T_1 = T_2$ and so $\Delta U = 0$. The first law then shows that $Q = W > 0$ where Q is the net heat supplied, and there is a 100% conversion of heat into work. However there is no violation of the second law here, because there has been a net change in the state of the ideal gas working system.

b *No effect other than* says that in addition to the rejection of heat to a body at a lower temperature, the only other effect on the surroundings is via the work delivered by the engine. This means that the bodies delivering and accepting heat to and from the engine must do so without delivering any work. In other words, their volumes must remain constant if only P-V work is being considered. Such a body that delivers its heat with no work is sometimes called a *source of heat*.

c Heat must be extracted from a *single* body. Suppose that heat $Q_1 + Q_2$ was supplied from two bodies: Q_1 from a body at T_1 and Q_2 from a body at T_2 with $T_1 > T_2$. The cyclical engine delivers an amount of work $W = Q_1 + Q_2$ as shown in Figure 4.5, and there appears to be a complete conversion of heat into work in a cyclical process with no heat being rejected to a reservoir at a lower temperature.

However, there is no violation of the second law here, because Q_2 could be negative with $W = |Q_1| - |Q_2|$. This possible type of engine is excluded from the Kelvin statement by specifying a single body.

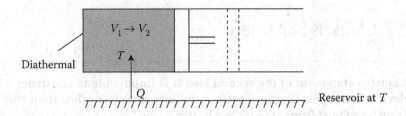

Figure 4.4   Isothermal expansion of an ideal gas. Although $W = Q$, this does not violate the second law.

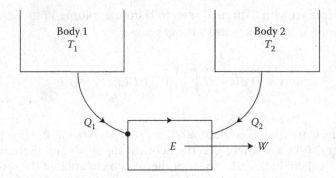

Figure 4.5   Two bodies supplying heat to an engine do not violate the Kelvin statement of the second law, because Q_2 could be negative.

It should be remarked here that, were one to run an engine with finite bodies as the sources of heat rather than reservoirs, it would eventually run down as the bodies approach each other in temperature. The result would be a less useful engine; however, the second law as given above applies to both situations.

The Kelvin statement of the second law is sometimes given in the very concise form:

A process whose only effect is the complete conversion of heat into work is impossible.

A few moments thought will show that the word *only* ensures that all the points made earlier are covered. The Kelvin statement was presented in the first more extended form, because its significance in the concise form is too easily overlooked.

4.3.2 Clausius Statement

Clausius statement of the second law: It is impossible to construct a device that, operating in a cycle, produces no effect other than the transfer of heat from a colder to a hotter body.

Schematically, this statement is represented in Figure 4.3(b). What this form of the second law tells us is that work must be performed if heat is to be transferred from a colder to a hotter body. Were this not so, you could heat your house just by cooling the outside air at no cost, with no work having to be done. Further, heat extracted from a colder body could be used to run a heat engine, and this would result in another *perpetuum mobile*.

A refrigerator, designated by R in Figure 4.3(b), is an engine that extracts heat from a cold body and delivers heat to a hot body when work is performed on the engine. The operation and efficiency of real refrigerators will be discussed in Section 4.6.

The hypothetical refrigerator in Figure 4.3(b) is not allowed because there is no external source of energy supplied to R. A real refrigerator depends on such energy and therefore is allowed by the Clausius statement of the second law.

There is one final point that should be discussed. The Kelvin statement of the second law refers to the impossibility of heat being extracted from a hot body and the performance of an equivalent amount of work, with there being no net change in the state of the working system. It does not forbid the opposite situation depicted in Figure 4.6, where all the work W done on an unchanged system may be converted completely into heat. Rumford's experiment with a blunt boring tool is an example of such a total conversion of work into heat. Another example is the operation of a resistive electrical circuit in which only work is performed, as in Figure 3.2(a).

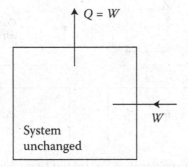

Figure 4.6 Work can be converted completely into heat, with no change in the working substance.

4.3.3 Equivalence of Kelvin and Clausius Statements

The two statements of the second law of thermodynamics are shown to be equivalent by showing that the falsity of each implies the falsity of the other.

Suppose first that the Kelvin statement is untrue. This means that one can have an engine E that takes Q_1 from a hot body and delivers work $W = Q_1$ in one cycle. Let this engine drive a refrigerator R as shown in Figure 4.7(a). Now adjust the size of the working cycles so that W is sufficient work to drive the refrigerator through one cycle. Suppose the refrigerator extracts heat Q_2 from the cold body. Then the heat delivered by it to the hot body is $Q_2 + W$ or $Q_1 + Q_2$. It is useful to regard the engine and the refrigerator as the composite engine enclosed by the dashed line as shown in Figure 4.7(b). This composite engine (strictly a refrigerator) extracts Q_2 from the cold body and delivers a net amount of heat $Q_2 + Q_1 - Q_1 = Q_2$ to the hot body, but no work is done. Hence, there is a violation of the Clausius statement.

Suppose now that the Clausius statement is untrue. This means that you can have a refrigerator that extracts heat Q_2 from a cold body and delivers the same heat Q_2 to a hot body in one cycle, with no work having to be done. Now let an engine operate between the same two bodies. Adjust the size of its working cycle so that in one cycle it extracts heat Q_1 from the hot body, gives up

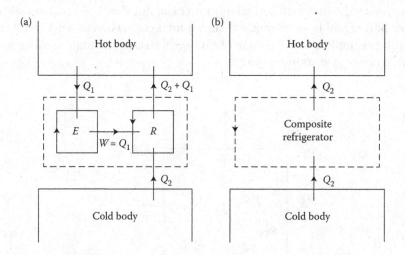

Figure 4.7 If the Kelvin statement of the second law is false, this implies that the Clausius statement is also false. The arrangement illustrated here is used to prove this.

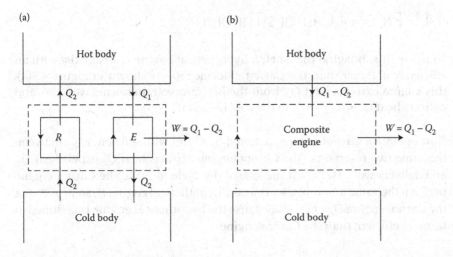

Figure 4.8   If the Clausius statement of the second law is false, this implies that the Kelvin statement is also false. The arrangement illustrated here is used to prove this.

the same heat Q_2 to the cold body as was extracted by the refrigerator, and so delivers the work $W = Q_1 - Q_2$. This is depicted in Figure 4.8(a). The engine and the refrigerator may be regarded as the composite engine enclosed by the dashed line, as shown in Figure 4.8(b), which takes in heat $Q_1 - Q_2$ from the hot body and delivers the same amount of work. Hence, there is a violation of the Kelvin statement. This proves the equivalence of the two statements.

4.4 CARNOT'S THEOREM

In the introduction to this chapter, you saw that Carnot had argued that efficient engines must be those operating as near as possible to a Carnot cycle. The Clausius statement of the second law leads directly to a theorem that relates the Carnot cycle to other heat-engine cycles.

> **Carnot's theorem: No engine operating between two reservoirs can be more efficient than a Carnot engine operating between those same two reservoirs.**

4.4.1 Proof of Carnot's Theorem

To prove this, imagine that such a hypothetical engine E' *does* exist with an efficiency η' larger than the Carnot efficiency η_C. As shown in Figure 4.9(a), this engine extracts heat Q'_1 from the hot reservoir, performs work W', and delivers heat $Q'_2 = (Q'_1 - W')$ to the cold reservoir.

Now operate a Carnot engine, denoted by C and with efficiency η_C, between the same two reservoirs. The Carnot engine extracts heat Q_1, expels heat Q_2, and delivers work W. Adjust the size of the cycle to make the Carnot engine perform the same amount of work as the hypothetical engine E', so $W' = W$. For the Carnot engine $Q_2 = Q_1 - W$. Because the hypothetical engine is assumed to be more efficient than the Carnot engine,

$$\frac{W'}{Q'_1} > \frac{W}{Q_1}$$

But $W' = W$, so

$$Q_1 > Q'_1$$

(a)   (b)

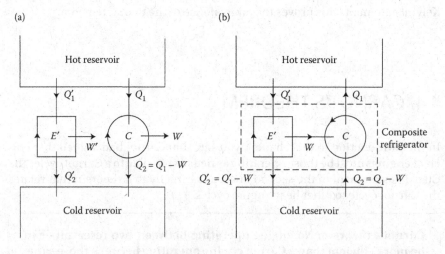

Figure 4.9   Arrangement used to prove Carnot's theorem: No engine working between two reservoirs can be more efficient that a Carnot engine working between the same two reservoirs.

A Carnot engine is a reversible engine, so it may be driven backwards as a refrigerator as shown in Figure 4.9(b). The hypothetical engine and the Carnot refrigerator together act as a composite device, shown by the dashed line, which extracts positive heat $(Q_1 - Q'_1)$ from the cold reservoir and delivers the same heat to the hot reservoir with no external work being required. But reservoirs are just large bodies in which the temperature is unchanged upon the addition of heat. This means that there is a violation of the Clausius statement. Therefore the engine E' cannot exist, and the original assumption that $\eta' > \eta_C$ is incorrect. It is permitted to have $\eta' = \eta_C$, at most. In that case the composite refrigerator simply transfers no net heat for no work, which is allowed.

We conclude that, for any real engine

$$\eta \le \eta_C$$

which proves the theorem.

4.4.2 Corollary to Carnot's Theorem

It follows from Carnot's theorem that:

> **All Carnot engines operating between the same two reservoirs have the same efficiency.**

To prove this statement, imagine two Carnot engines C and C' operating between the same two reservoirs, and let the size of the working cycles be adjusted so that they each deliver the same amount of work.

Let C run C' backwards as in Figure 4.10. It follows from the argument given in Section 4.4.1 that

$$\eta \le \eta'_C$$

If C' now runs C backwards,

$$\eta'_C \le \eta$$

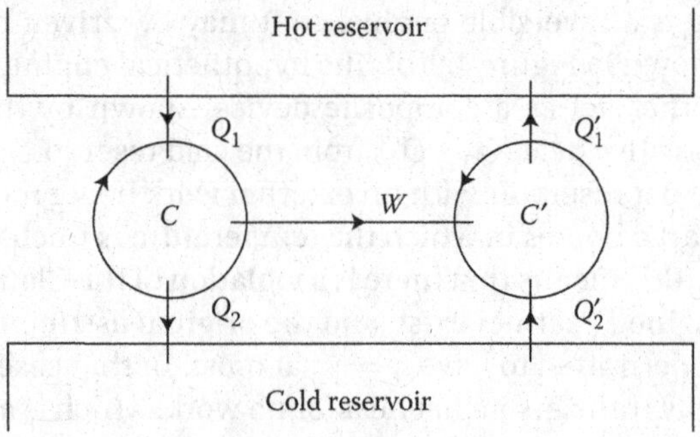

Figure 4.10 Arrangement used to prove that all Carnot engines operating between the same two reservoirs have the same efficiency.

Therefore,

$$\boxed{\eta_C = \eta_C'}$$

which proves the assertion.

4.5 THE THERMODYNAMIC TEMPERATURE SCALE

You have just seen that the efficiency of a Carnot engine operating between the same two reservoirs is independent of the working substance and can thus depend only on the temperatures of the reservoirs. This gives a means of defining a temperature scale that is independent of any particular material.

4.5.1 Temperature Scale from Carnot Engines

Define the thermodynamic temperature T so that T_1 and T_2 for the two reservoirs in a Carnot engine are related as

$$\eta_C = \frac{T_1 - T_2}{T_1} = 1 - \frac{T_2}{T_1} \qquad\qquad 4.2$$

Comparing this with Equation 4.1,

$$\frac{T_1}{T_2} = \frac{Q_1}{Q_2} \qquad \text{(Carnot)} \qquad 4.3$$

where the Carnot in parentheses emphasizes that this definition holds only for a Carnot engine. Note that consistent with Equation 4.1 the heat flow Q_1 in and the heat flow Q_2 out are both taken to be positive numbers.

Figure 4.11 helps illustrate why Equation 4.3 gives a sensible definition for a scale of temperature. The Carnot engine C_{12} operates between the reservoirs at T_1 and T_2. For this engine, Equation 4.3 gives

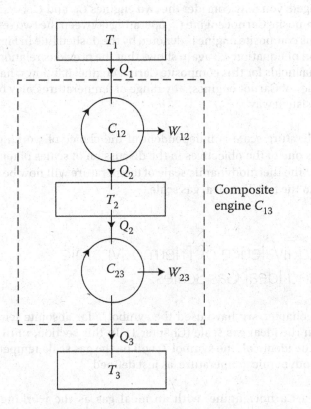

Figure 4.11 Thermodynamic temperature scale, as defined by Equation 4.3, is consistent with the arrangement illustrated here.

$$\frac{T_1}{T_2} = \frac{Q_1}{Q_2} \qquad\qquad 4.4$$

Suppose a second Carnot engine C_{23} operates between the reservoir at T_2 and a third reservoir at T_3. Let C_{23} absorb the same amount of heat Q_2 from the reservoir at T_2 as was rejected to that reservoir by C_{12}. When the two engines operate together, the reservoir at T_2 is thus unchanged. Equation 4.3 gives

$$\frac{T_2}{T_3} = \frac{Q_2}{Q_3} \qquad\qquad 4.5$$

Multiplying Equation 4.4 by Equation 4.5,

$$\frac{T_1}{T_3} = \frac{Q_1}{Q_3}$$

which does not involve the intermediate temperature T_2. As the reservoir at T_2 is unchanged, you may consider the two engines C_{12} and C_{23}, acting together, to be a composite Carnot engine C_{13} operating between the two reservoirs at T_1 and T_3. This composite engine is denoted by the dashed line in Figure 4.11. The application of Equation 4.3 again shows that the previous relation is precisely the one that holds for this composite Carnot engine. It follows that by taking a whole series of Carnot engines, any range of temperatures may be defined in a self-consistent way.

This temperature scale is independent of the choice of working substance, which was one of the objectives in the discussion of scales of temperature in Chapter 1. The thermodynamic scale of temperature will now be shown to be identical to the familiar ideal gas scale.

4.5.2 Equivalence of Thermodynamic and Ideal Gas Scales

Until this chapter, we have used the symbol T for absolute temperature as defined on the ideal gas scale (Chapter 1). In this section, until the two are proved to be identical, the symbol T_g will be the gas scale temperature and T the thermodynamic temperature, as just defined.

Consider a Carnot engine, with an ideal gas as the working substance, operating between the two reservoirs at the ideal gas scale temperatures T_{g_1} and T_{g_2}. Follow the operating cycle abcda shown in Figure 4.12. For the isotherm bc, the empirical equation of state involving the gas scale temperature T_{g_1} is

$$PV = nRT_{g_1} \qquad\qquad 4.6$$

The first law gives for an infinitesimal part of this reversible process

$$đQ = dU + PdV = PdV \qquad\qquad 4.7$$

where the last equality is true because the temperature is constant, making $dU = 0$. The heat Q_1 entering the engine in this portion of the cycle is

$$Q_1 = \int_{V_b}^{V_c} PdV = nRT_{g_1} \int_{V_b}^{V_c} \frac{1}{V}dV$$

$$= nRT_{g_1} \ln\left(\frac{V_c}{V_b}\right) \qquad\qquad 4.8$$

Q_1 is positive if $V_c > V_b$, which is consistent with the idea that heat enters the engine in this portion of the cycle. Similarly, the heat entering the engine along the da isotherm part of the cycle is $nRT_{g_2}\ln(V_a/V_d)$. This is negative if $V_a < V_d$, which means that heat flows out of the engine. However, in Figure 4.12 Q_2 has been defined as the heat flow out of the engine, so

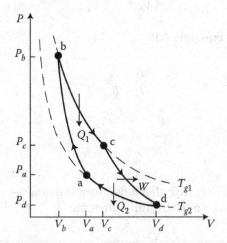

Figure 4.12 A Carnot cycle with an ideal gas as the working substance. This figure is used to show that the ideal gas scale temperature is identical to the thermodynamic temperature T.

$$Q_2 = -nRT_{g_2}\ln\left(\frac{V_a}{V_d}\right) = nRT_{g_2}\ln\left(\frac{V_d}{V_a}\right)$$ 4.9

Dividing Equation 4.8 by Equation 4.9,

$$\frac{Q_1}{Q_2} = \frac{T_1}{T_2} = \frac{T_{g_1}\ln(V_c/V_b)}{T_{g_2}\ln(V_d/V_a)}$$ 4.10

But ab and cd are adiabats where $T_g\,V^{\gamma-1}$ = constant, holds so:

$$T_{g_1}V_c^{\gamma-1} = T_{g_2}V_d^{\gamma-1} \quad (\text{cd adiabat})$$ 4.11

$$T_{g_1}V_b^{\gamma-1} = T_{g_2}V_a^{\gamma-1} \quad (\text{ab adiabat})$$ 4.12

Dividing Equation 4.11 by Equation 4.12,

$$\frac{V_c}{V_b} = \frac{V_d}{V_a}$$

or

$$\ln\left(\frac{V_c}{V_b}\right) = \ln\left(\frac{V_d}{V_a}\right)$$

Substituting this in Equation 4.10,

$$\frac{T_1}{T_2} = \frac{T_{g_1}}{T_{g_2}}$$

This means that

$$T_g = \varepsilon T$$

where ε is a constant. Because all temperature scales agree at the fixed point of 273.16 K, the constant ε must be unity. Therefore,

$$T_g \equiv T$$ 4.13

That is,

> **the thermodynamic and the ideal gas scales of temperature are identical.**

4.6 ENGINES AND REFRIGERATORS

An example of a real engine will be illustrated in Section 4.6.4, but it is instructive to think first about the efficiency of an engine based on the Carnot cycle. This has a theoretical use in that it gives an upper limit, by Carnot's theorem, for the efficiency of any possible engine that might be designed.

4.6.1 Efficiency of a Heat Engine

A Carnot engine, such as the one depicted in Figure 4.1, has efficiency

$$\eta_C = 1 - \frac{Q_2}{Q_1} = 1 - \frac{T_2}{T_1} \qquad\qquad 4.14$$

It is a simple matter to calculate the efficiency, knowing T_1 and T_2. For example, consider a steam engine that operates between a maximum steam temperature of 500°C (or 773 K) and an ambient temperature of 20°C (or 293 K). Then by Equation 4.14 the Carnot efficiency is

$$\eta_C = 1 - \frac{T_2}{T_1} = 1 - \frac{293K}{773K} = 0.62$$

Real steam engines, such as in a fossil-fuel driven power plant, typically have efficiencies of about 0.4. The Carnot efficiency is a significant overestimate of efficiency for most real engines.

The reason for this difference is that a real steam engine (or any practical engine) operates nothing like the Carnot cycle described in Section 4.1. In a real heat engine, there is no attempt to set up the isothermal and adiabatic steps shown in Figure 4.1. Further, any device that operates at a high temperature invariably loses energy to its environment, despite efforts to provide insulation barriers.

It is interesting to note that the Carnot engine's efficiency would be 100% were it possible to obtain a lower temperature reservoir at absolute zero. However, this is forbidden by the third law (Chapter 12). One might argue then that a good strategy for increasing efficiency would be to use a refrigeration system to create a low-temperature reservoir with much less than ambient temperature, say 100 K or lower. The counterargument against this strategy is that there is an energy cost to refrigerate, a cost that as you will see in Section 4.6.2 increases as the refrigeration temperature gets lower. This offsets any gain obtained by increasing a heat engine's theoretical efficiency.

> **Although the Carnot efficiency η_C is a useful theoretical tool, a Carnot engine would never be practical for delivering work. Apart from the difficulty in setting up the required isothermal and adiabatic steps, heat flow in an isothermal process is painfully slow. If you want a machine that generates significant electrical energy or powers a vehicle of some kind, energy must be available at a desirably high rate. This is beyond the ability of a Carnot engine.**

4.6.2 Refrigerators and COP

Imagine now that the Carnot engine is run backwards, as in Figure 4.13, to act as a refrigerator. In keeping with the concept of efficiency as a ratio of effective output to input, the "efficiency" η_C^R of a refrigerator is the heat extracted from the cold reservoir divided by the work input. So as not to confuse this with the efficiency of a heat engine, it is customary to drop the word efficiency and instead refer to η_C^R as the *coefficient of performance*, or simply COP. For a Carnot refrigerator,

$$\eta_C^R = \frac{Q_2}{W} = \frac{Q_2}{Q_1 - Q_2} = \frac{T_2}{T_1 - T_2}$$ 4.15

In this case W is the energy input into the device, normally electrical energy for a household refrigerator or air conditioner.

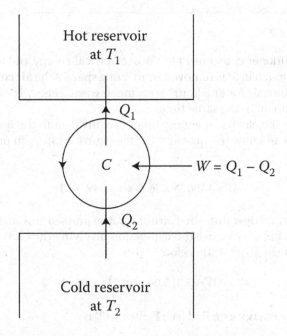

Figure 4.13 A Carnot refrigerator.

Commercially produced refrigerators now typically come with a COP rating posted in their specifications.

For example, a household refrigerator in a 20°C room (293 K) might maintain a refrigerator compartment at 3°C (276 K). For this system the Carnot model predicts a maximum COP η_C^R of about 16. However, for the freezer compartment maintained at −18°C (255 K), the maximum COP is only about 6.7. Real household refrigerators have coefficients of performance of about 4 or 5, which, as expected, is less than the theoretical maximum.

At first it may appear strange to have the COP, a type of efficiency rating, be larger than 1. Rest assured that this does not violate any laws of thermodynamics. The COP is simply a measure of how effectively the input energy W moves thermal energy from the cold reservoir at T_2 to the hot reservoir at T_1. Notice that the theoretical COP of any refrigerator drops as the temperature difference increases.

EXAMPLE 4.2

An air conditioner consumes 1400 W of electrical energy and has a COP of 3.3. How much heat is removed from living space if the air conditioner runs continuously for one hour? How much waste energy is expelled to the environment in the same time?

Solution: The electrical energy input is equivalent to the quantity W, and we seek to know the quantity Q_1 (see Figure 4.13). With one hour = 3600 s,

$$W = 1400 \text{ W} \times 3600 \text{ s} = 5.04 \times 10^6 \text{ J}$$

For a real refrigerator, the Carnot limit expressed in Equation 4.15 is not valid, but we can use the expression involving the COP and heat, which is true in general. Therefore

$$Q_2 = \eta^R W = 3.3\left(5.04 \times 10^6 \text{ J}\right) = 1.66 \times 10^7 \text{ J}$$

The waste energy expelled (see Figure 4.13) is

$$Q_1 = Q_2 + W = 1.66 \times 10^7 \text{ J} + 5.04 \times 10^6 \text{ J} = 2.16 \times 10^7 \text{ J}$$

Notice that this real air conditioner does not approach the Carnot COP. If say T_2 (room temperature) = 22°C = 295 K and the outdoor temperature is T_2 = 35°C = 308 K, the Carnot COP would be almost 24, which is far higher than the real air conditioner.

4.6.3 Heat Pumps

In a refrigerator or air conditioner, Q_1 is the waste heat delivered to the environment. You can easily feel the warm air expelled from a refrigerator. But in an important application called a *heat pump*, Q_1 is not wasted but rather used to maintain a warm environment, for example heating a home on a cold day.

For a heat pump the schematic cycle is the same as for a refrigerator (Figure 4.13), but now the effective output is the heat delivered Q_1. Therefore the COP becomes $\eta^{HP} = Q_1/W$, which in the Carnot limit is

$$\eta_C^{HP} = \frac{Q_1}{W} = \frac{Q_1}{Q_1 - Q_2} = \frac{T_1}{T_1 - T_2} = \frac{1}{1 - T_2/T_1} \qquad 4.16$$

For example, on a cold day with outdoor air temperature $T_2 = 0°C$ (273 K) and indoor temperature $T_1 = 20°C$ (293 K), the Carnot COP is $\eta_C^{HP} = 15$. The COP for a real heat pump under these conditions is perhaps 3 or 4, which is still much better than using a traditional heat source (see Problem 4.10). Another advantage of having a heat pump is that in summer it can be turned around to cool the interior air, so no separate air conditioning unit is required.

Figure 4.14 shows the efficiency of a Carnot heat pump graphed against the temperature ratio T_2/T_1. Just as with a refrigerator, the efficiency drops as the temperature difference increases (or ratio decreases). Therefore, heat pumps are less attractive in climates that are subject to extreme temperatures, where an additional conventional heat source might be needed as a supplement.

4.6.4 Internal Combustion Engines and the Otto Cycle

The Carnot engine is an idealized engine. Real engines operate in various cycles, all different from the idealized Carnot one. An important example historically as well as presently is the internal combustion engine, the traditional gasoline-powered engine used in automobiles and other applications.

We present two versions of the PV diagram for this engine cycle. Figure 4.15(a) is a fairly realistic representation of the actual engine cycle. These are the key steps in the process:

1. Gasoline vapor and air are drawn into the cylinder. This is the nearly horizontal line at the bottom of the diagram, leading to the point labeled a.

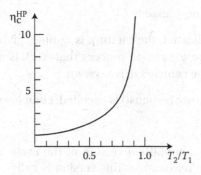

Figure 4.14 "Efficiency" of a Carnot heat pump as a function of the ratio of the reservoir temperatures.

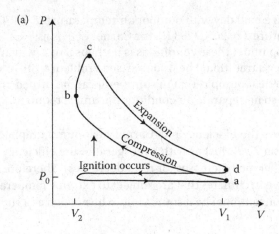

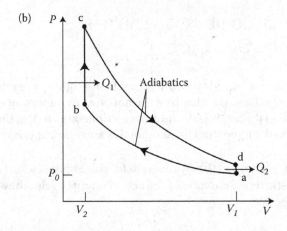

Figure 4.15 Four-stroke internal combustion engine: (a) the actual cycle and (b) the Otto cycle, an idealized representation of the actual cycle.

2. The mixture is compressed.

3. At the point indicated, the mixture is ignited, giving rise to the power stroke. It's in the expansion process that work is done and eventually transferred to the vehicle's drive system.

4. The burned mixture (exhaust) is expelled, completing the cycle.

Figure 4.15(b) shows a simplified version of the cycle, which makes quantitative analysis more reasonable. This version is called the *Otto cycle*, after German engineer Nikolaus August Otto, who made a practical internal combustion engine in 1867. Rather than trying to deal with changing gas mixtures throughout the cycle, the gas is simply assumed to be air, which itself is assumed to be an ideal, diatomic gas with adiabatic exponent $\gamma = 7/5$. Both the compression and expansion of the cylinder are assumed to take place adiabatically, which is not a bad assumption, because the cycle happens so quickly. Ignition takes place during step bc, with a drastic increase in pressure at constant volume V_2. Exhaust is done in step da at constant volume V_1. With compression and expansion assumed to be adiabatic, heat Q_1 enters in ignition and heat Q_2 exits in exhaust, as shown.

The following is an assessment of the thermodynamics of each step of the Otto cycle.

a–b The piston moves to compress the gas reversibly and adiabatically with

$$T_a V_1^{\gamma-1} = T_b V_2^{\gamma-1} \qquad 4.17$$

b–c Heat Q_1 is added at constant volume from an external source with

$$Q_1 = C_V (T_c - T_b) \qquad 4.18$$

c–d The gas expands adiabatically and reversibly in the power stroke with

$$T_d V_1^{\gamma-1} = T_c V_2^{\gamma-1} \qquad 4.19$$

d–a At the bottom of the power stroke, the gas is assumed to cool at constant volume to pressure P_0 by giving up heat Q_2 to external reservoirs, with

$$Q_2 = C_V (T_d - T_a) \qquad 4.20$$

It is possible to derive an expression for the efficiency using Equations 4.1, 4.18, and 4.20. For this cycle,

$$\eta = 1 - Q_2/Q_1 = 1 - \left[\frac{T_d - T_a}{T_c - T_b} \right]$$

Equations 4.17 and 4.19 for the adiabatic processes give, on subtraction,

$$(T_d - T_a) V_1^{\gamma-1} = (T_c - T_b) V_2^{\gamma-1}$$

or

$$\left(\frac{V_1}{V_2} \right)^{\gamma-1} = \frac{T_c - T_b}{T_d - T_a}$$

For obvious reasons, the ratio V_1/V_2 is known as the *compression ratio* r_c, so

$$\eta = 1 - \left(\frac{V_2}{V_1}\right)^{\gamma-1} = 1 - \frac{1}{r_c^{\gamma-1}} \qquad 4.21$$

Obviously, efficiency is improved with a higher compression ratio. In automobiles, values for r_c of about 7 or 8 are typical, giving a theoretical efficiency of $1 - 1/7^{0.4} = 0.54$. The actual efficiency of a real internal combustion engine is much lower than this idealized value, normally about 30%. High-performance cars are designed to have r_c of 10 or more, approaching $r_c = 17$ for Formula 1 racing cars.

The internal combustion engine is, relative to other alternatives, a very efficient one, which explains why it has been popular for use in automobiles for over 100 years. Its main disadvantages are that it relies on non-renewable fossil fuel and emits hydrocarbons that act as greenhouse gases. Alternative energy sources for automobiles include the fuel cells discussed in Chapter 7. As of the early 21st century, the most popular alternative is the hybrid automobile, which harnesses together an internal combustion engine with rechargeable batteries.

4.7 SUMMARY

This has been an important chapter, and it is useful to summarize the results, which are not only useful by themselves but also in developing the concept of entropy (Chapter 5).

1. A heat engine converts heat into work in a cyclical process in which the working substance is unchanged.

2. A Carnot engine is a reversible engine that operates between two temperatures only. In general, engines take in and reject heat at a variety of temperatures.

3. The efficiency of a heat engine is

$$\eta = 1 - \frac{Q_2}{Q_1}$$

4. The essence of the Kelvin statement of the second law is that a cyclical engine cannot convert heat from a single body at a uniform temperature completely into work. Some heat has to be rejected at a lower temperature. The essence of the Clausius statement is that heat cannot flow from a cold body to a hot body by itself—work has to be done in a cyclical refrigerator to achieve this.

5. The most efficient engine operating between a given pair of reservoirs is a Carnot engine. All Carnot engines operating between the same reservoirs have the same efficiency, independent of the nature of the working substance.

6. For a Carnot engine, the thermodynamic temperature is defined to be

$$\frac{Q_1}{Q_2} = \frac{T_1}{T_2} \qquad \text{with} \qquad \eta_C = 1 - \frac{T_2}{T_1}$$

The thermodynamic temperature is identical to the ideal gas temperature.

Problems

4.1 Heat is supplied to an engine at the rate of 10^6 J/min, and the engine has a rated output of 10 horsepower. (a) What is the efficiency of the engine? (b) What is the heat output per minute?

4.2 A storage battery delivers a current into an external circuit and performs electrical work. The battery remains at a constant temperature by absorbing heat from the surrounding atmosphere. Heat then appears to be completely converted into work. Is this a violation of the second law? Explain.

4.3 Show that two adiabatic lines on a PV diagram cannot intersect. (Hint: Imagine that they do; complete a cycle with an isotherm and operate an engine around this cycle.)

4.4 An inventor claims to have developed an engine that takes in 1.1×10^8 J at 400 K, rejects 5.0×10^7 J at 200 K, and delivers 16.7 kW hours of work. Would you advise investing money in this project?

4.5   Which gives the greater increase in the efficiency of a Carnot engine: increasing the temperature of the hot reservoir or lowering the temperature of the cold reservoir by the same amount?

4.6   Consider again the three-step cycle described in Chapter 3, Problem 3.14. (a) Find the efficiency of a heat engine that operates using this cycle. (b) Find the Carnot efficiency of this engine and compare the result to the real efficiency calculated in (a). Discuss the effectiveness of this engine.

4.7   An electrical generating plant produces energy at a rate of 1.5 GW with an efficiency of 0.35. (a) Find the energy needed to run this plant and the waste heat discarded, both as rates in GW. (b) If the waste heat is dumped into the environment which is at 25°C, what is the minimum boiler temperature?

4.8   Fifty kg of liquid water initially at 0°C is frozen into ice in a refrigerator. The room temperature is 20°C. What is the minimum work input to the refrigerator to achieve this? (Latent heat of fusion of water = 3.336×10^5 J/kg.)

4.9   It is proposed to heat a house using a heat pump operating between the house and the outside. The house is to be kept at 22°C, the outside is at –10°C, and the heat loss from the house is 15 kW. What is the minimum power required to operate the pump?

4.10 Suppose a house requires 4.3 GJ of heating in a winter month. The utility company charges \$0.14 per kWh. (a) Find the cost savings of using a heat pump versus a 95%-efficient natural gas furnace. Assume a Carnot heat pump with average temperatures of 20°C indoors and 0°C outdoors. (b) Repeat part (a) using a more realistic coefficient of performance of 4.0 for the heat pump.

4.11 In low-temperature physics, a common refrigerant is liquid nitrogen, with a temperature of 77 K at $P = 1$ atm. (a) What is the maximum coefficient of performance of a refrigerator designed to maintain that temperature inside a lab at 20°C? (b) For work at extremely low temperatures, liquid helium with a boiling point of 4.2 K is used. Repeat part (a) for a refrigerator that maintains this temperature.

4.12 Show that the efficiencies of the three Carnot engines, operating between the three reservoirs as illustrated in Figure 4.16, are related by

$$\eta_3 = \eta_1 + \eta_2 - \eta_1\eta_2$$

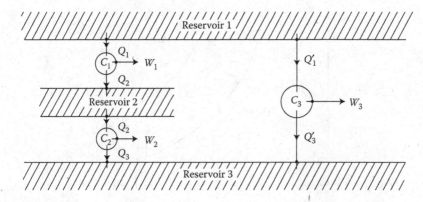

Figure 4.16

4.13 A Carnot engine working on a satellite in outer space has to deliver a fixed amount of power at rate W. The temperature of the heat source is also fixed, at T_1. The lower temperature reservoir at T_2 consists of a large body of area A; its temperature is maintained at T_2 because it radiates energy into space as much heat as is delivered to it by the engine. The rate of this radiation is $\sigma A T_2^4$ where σ is a constant. The Carnot engine has to be designed so that, for a given W and T_1, A has a minimum value. Show that A has a minimum value when T_2 takes the value $3T_1/4$.

4.14 A hypothetical engine, with an ideal gas as the working substance, operates in the cycle shown in Figure 4.17. Show that the efficiency of the engine is

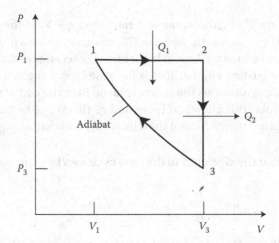

Figure 4.17

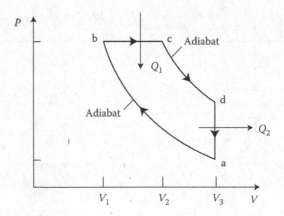

Figure 4.18

$$\eta = 1 - \frac{1}{\gamma}\left(\frac{1 - P_3/P_1}{1 - V_1/V_3}\right)$$

4.15 A simplified representation of the Diesel cycle, with just air as the working substance, is as shown in Figure 4.18. Show that the efficiency of this engine is

$$\eta = 1 - \frac{1}{\gamma}\left(\frac{\left(1/r_e^{\gamma}\right) - \left(1/r_c^{\gamma}\right)}{\left(1/r_e\right) - \left(1/r_c\right)}\right)$$

where $r_e = V_3/V_2$, the expansion ratio, and $r_c = V_3/V$, the compression ratio. If $r_e = 5$, $r_c = 15$, and $\gamma = 7/5$, evaluate η. Notice that the compression ratio can be much higher in a Diesel engine than in an internal combustion engine. That is because Diesel engines do not suffer from pre-ignition, as the fuel is sprayed in at the end of the compression stroke; this allows an increased r_c. This is one reason why Diesel engines are more efficient than internal combustion engines.

4.16 Show that the efficiency of the Otto cycle can be expressed as either

$$\eta = 1 - \frac{T_a}{T_b} \qquad \text{or} \qquad \eta = 1 - \frac{T_d}{T_c}$$

Show that both of these results are lower than the Carnot efficiency.

4.17 (a) Discuss whether the direction of travel around the PV-diagram curve should be clockwise or counterclockwise for a heat engine. (b) Discuss whether the direction of travel around the PV-diagram curve should be clockwise or counterclockwise for a refrigerator.

4.18 A power plant with efficiency 0.32 produces electrical energy at a steady rate of 1000 MW. The waste heat is all dumped into a point in the river where the flow rate is 1250 m^3/s. What is the water temperature increase due to the waste heat?

4.19 A large building is heated in winter by a heat pump that has a COP of 4.0 and supplies heat at a rate of 90 kW. How much energy is used in one day to heat the building? How much does it cost if the incoming energy costs $0.15 per kWh?

BIBLIOGRAPHY

Curzon, F.L. and Ahlborn, B., Efficiency of a Carnot engine at maximum power output, *American Journal of Physics 43*, 22–24, 1975.

Kittel, C. and Kroemer, H., *Thermal Physics*, second edition, W.H. Freeman, New York, 1980.

Chapter 5: Entropy

The concept of entropy is fundamental to many aspects of thermodynamics. In this chapter we define and develop the idea of entropy and provide some important examples. In Chapter 6 and beyond, entropy is understood from a statistical perspective as well as a thermodynamic one.

5.1 THE CLAUSIUS INEQUALITY

As a prelude to discussing entropy, there is an important theorem that applies to cyclical processes. This theorem is known as the Clausius inequality.

5.1.1 Development of the Clausius Inequality

Consider a working substance undergoing a cycle so that, at the end of the cycle, its state is unchanged. In Figure 5.1(a) this cycle is represented symbolically by the circle in the center. The starting state is at the temperature T_1 and is represented by the point 1. The engine is driven by a principal reservoir at temperature T_0, assumed to be large enough that its temperature does not change appreciably when it supplies heat to the engine.

The working substance in the engine is driven around a cycle in the following way. The state of the working substance is first changed to an infinitesimally close neighboring state 2 at temperature T_2 by injecting a small amount of heat δQ_1. This is done with a Carnot engine C_1, which operates between two auxiliary reservoirs at T_0 and T_1. The auxiliary reservoir at T_1 supplies heat δQ_1 to the working substance, and an equal quantity of heat is supplied by C_1 to that reservoir to leave it unchanged. C_1 in turn takes heat $(T_0/T_1)\,\delta Q_1$ from the auxiliary reservoir at T_0 and performs work δW_1. If the auxiliary reservoir at T_0 is to remain unchanged, heat $(T_0/T_1)\,\delta Q_1$ enters it from the principal reservoir. In this way, the change from 1 to 2 is made with the only other changes (a) the

DOI: 10.1201/9781003299479-5

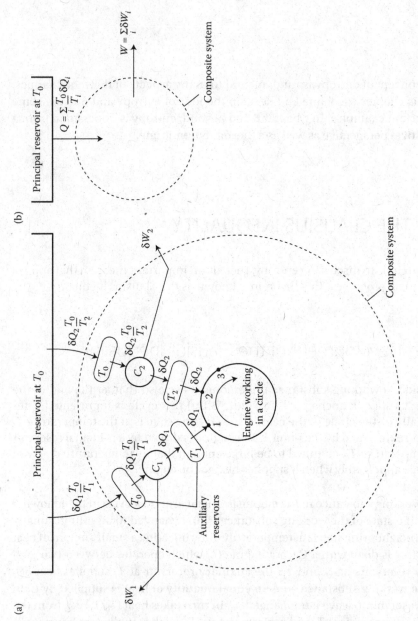

Figure 5.1   Cycle used to derive the Clausius inequality.

performance of the external work δW_1 and (b) the extraction of heat $(T_0/T_1)\delta Q_1$ from the principal reservoir.

The process is repeated, taking the working substance from 2 to 3 with the help of the Carnot engine C_2 and a new pair of auxiliary reservoirs at T_2 and T_0, and so on to complete the cycle.

Consider now the composite system consisting of the working system, all the Carnot engines, and all the auxiliary reservoirs. This composite system includes everything within the dashed line in Figure 5.1(a). At the end of the cycle

1. Everything in the composite system is unchanged, and so $\Delta U = 0$.

2. The heat supplied to the composite system is

$$Q = \sum_i \delta Q_i \frac{T_0}{T_i}$$

where the summation is over all the Carnot engines used.

3   The external work performed is

$$\sum_i \delta W_i \equiv W$$

Applying the first law to the composite system,

$$0 = Q - W \qquad \text{or} \qquad W = Q$$

This situation is represented in Figure 5.1(b), where it is shown that heat has been extracted from a single reservoir and used to perform an equal amount of work. This is a violation of the Kelvin statement of the second law. The only way this process can occur is for both W and Q to be negative; that is, work is done *on* the system and an equal quantity of heat flows out. This is just the allowed situation of Figure 4.6. Alternatively, both W and Q could be zero. Therefore, one may conclude that

$$W = Q \leq 0$$

From the analysis above, this means that

$$T_0 \sum_i \frac{\delta Q_i}{T_i} \leq 0 \quad \text{so} \quad \sum_i \frac{\delta Q_i}{T_i} \leq 0$$

In the limit of infinitesimal changes,

$$\oint \frac{dQ}{T} \leq 0$$

where the circle on the integral sign indicates that the cycle is complete or closed. This is known as the *Clausius inequality* and is one of the key results in thermodynamics.

5.1.2 Discussion and Implications

Before moving on, three important points should be made.

1.   The proof of the inequality emphasizes that the T appearing inside the integral is the temperature of the auxiliary reservoirs supplying heat to the working substance. It is thus the temperature of the external source of heat. The Clausius inequality is written as

$$\oint \frac{dQ}{T_0} \leq 0 \qquad \text{(Clausius inequality)} \qquad \text{5.1}$$

where T_0 is written to remind us of this.

2.   If the cycle is reversible so the infinitesimal heat flow and work done each have the same magnitude when running in reverse, the proof would give

$$\oint \frac{dQ}{T_0} \geq 0$$

W would then be done on the composite system, with an equal amount of heat $T_0 \sum_i \delta Q_i / T_i$ being rejected to the principal reservoir. This does not violate the Kelvin statement, providing

$$W = Q = T_0 \sum_i \frac{\delta Q_i}{T_i} \geq 0$$

The only way for both inequalities to be satisfied is for

$$\oint_R \frac{dQ_R}{T} = 0 \qquad (\text{reversible cycle only}) \qquad \text{5.2}$$

The placement of R at the bottom of the integral sign and as a subscript to dQ emphasizes that this relation is valid only for a reversible process. However, the 0 subscript on T has been dropped, because there is now no difference between the temperature of the external source supplying the heat and the temperature of the working substance.

3. The sign of the inequality follows from the fact that, in the proof, heat was always flowing into the engine. This requires that $T_0 > T$ and that the equality sign holds for the reversible case where $T_0 = T$. Replacing T by the larger T_0 makes the inequality less than zero.

5.2 ENTROPY

5.2.1 Definition

This concept follows immediately from Section 5.1. Suppose a system is taken along a reversible path R_1 from an initial state i to a final state f and then back again to the initial state along another reversible path R_2, completing a reversible cycle. Figure 5.2 illustrates this for a gas system.

Because the cycle is reversible, the equality sign holds in the Clausius inequality (Equation 5.2). Remembering that the cycle is composed of the two reversible paths R_1 and R_2,

$$\oint_R \frac{Q_R}{T} = \int_{R_1 f}^{i} \frac{Q_R}{T} + \int_{R_2 i}^{f} \frac{dQ_R}{T} = 0$$

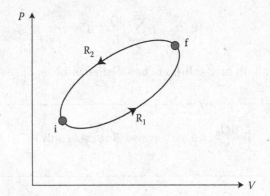

Figure 5.2   A reversible cycle. The text shows that $\int dQ/T$ is the same for the reversible paths R_1 and R_2 connecting i and f.

so

$$\int_{R_1 i}^{f} \frac{dQ_R}{T} = - \int_{R_2 f}^{i} \frac{dQ_R}{T}$$

But

$$\int_{R_2 i}^{f} \frac{dQ_R}{T} = - \int_{R_2 f}^{i} \frac{dQ_R}{T}$$

because R_2 is reversible. Thus,

$$\int_{R_1 i}^{f} \frac{dQ_R}{T} = \int_{R_2 i}^{f} \frac{dQ_R}{T}$$

which means that the integral $_R\int_i^f (dQ/T)$ is path independent. This means that there must be a state function S with

$$\boxed{\Delta S = S_f - S_i = \int_{R\,i}^{f} \frac{dQ_R}{T}}$$ 5.3

This state function is called *entropy*. Notice that only entropy differences have been defined. Also, it cannot be stressed too strongly that the defining integral for entropy differences has to be taken over a reversible path. To summarize:

For an infinitesimal reversible process:

$$\dbar Q_R = T\, dS\,(\text{reversible only}) \qquad\qquad 5.4$$

The name entropy comes from the Greek *en* meaning inside and *tropos* meaning transformation. Clausius invented the word (die Entropie in German) in 1865. He intended the word to convey the idea of heat being converted into work in an engine.

5.2.2 Example of Entropy Change in Water

As a first example, consider a beaker of water at atmospheric pressure that is heated from 20°C to 100°C by placing it in thermal contact with a reservoir at 100°C. When the water reaches 100°C, the beaker is removed from the reservoir and placed in an insulating jacket. The process is shown in Figure 5.3. Heat passes from the reservoir into the water, and it might seem that a simple application of Equation 5.3 would suffice. However this equation applies to a reversible process, while the actual process here is irreversible because of the inherent finite temperature differences.

Consider again the argument encountered in Section 2.1.4. The water is in initial and final equilibrium states, each with well-defined entropies. Thus, the

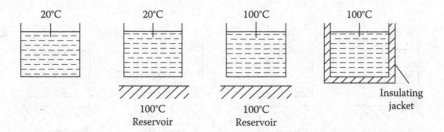

Figure 5.3 A beaker of water is heated irreversibly and isobarically between 20°C and 100°C.

entropy change for this process is also well defined. One may simply imagine any convenient reversible process that takes the system between the same two end points and calculate the entropy change for this imaginary process, using Equation 5.3. This entropy change is then the same as that occurring in the actual irreversible process.

One simple reversible heating process between the end points could be affected by bringing up a whole series of reservoirs between 20°C and 100°C, keeping the pressure constant, so that the water passes through a series of equilibrium states. This process is shown in Figure 5.4.

When the water is at T and it is heated to $T + dT$ by thermal contact with the reservoir at $T + dT$, the heat entering the water reversibly is

$$đQ_R = C_P dT$$

where C_p is the water's heat capacity at constant pressure. Hence, the entropy change of the water is given by Equation 5.4 as

$$dS = \frac{đQ_R}{T} = \frac{C_P}{T} dT$$

or

$$\Delta S = C_P \int_{T_i}^{T_f} \frac{1}{T} dT = C_P \ln\left(\frac{T_f}{T_i}\right) \qquad 5.5$$

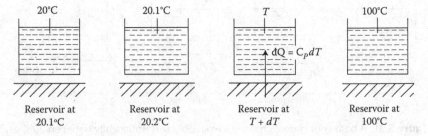

Figure 5.4   The same beaker of water is heated reversibly and isobarically between 20°C and 100°C in an imaginary process.

To compute the entropy change, remember that the temperatures here represent the thermodynamic temperature T. For example, for exactly 1 kg of water with $C_P = 4.19$ kJ/K,

$$\Delta S = (4190 \text{ J/K}) \ln\left(\frac{373\text{K}}{293\text{K}}\right) = 1.01 \times 10^3 \text{ J/K}$$

Any other reversible path would give, of course, the same answer, but this path is probably the most convenient.

> **Notice that absolute temperature (in K) must be used in this computation, and the SI units of entropy are J/K.**

5.2.3 Entropy Change in Free Expansion of an Ideal Gas

As a second example of an entropy change, consider an ideal gas undergoing a free expansion doubling its volume (see Sections 2.2.2 and 3.5.1).

In a free expansion (i) the process is irreversible; (ii) there is no temperature change; and (iii) no heat enters the system, because the walls are adiabatic. In order to apply Equation 5.3 to find the entropy change, imagine a reversible isothermal doubling of the volume. Such an expansion could be achieved by allowing the gas to expand slowly while in thermal contact with a reservoir at T. Applying the first law to this process:

$$đQ = dU + PdV = PdV$$

where $dU = 0$ because T is constant. Thus,

$$dS = \frac{P}{T}dV = \frac{nR}{V}dV$$

where the last step uses the equation of state $PV = nRT$. Then

$$\Delta S = nR \int_{V}^{2V} \frac{1}{V} dV = nR\ln 2 \qquad\qquad 5.6$$

This is the entropy change in an irreversible free expansion with a doubling of volume. For example, for one mole of an ideal gas

$$\Delta S = nR \ln 2 = (1 \text{ mol}) \left(\frac{8.315 \text{ J}}{(\text{mol} \cdot \text{K})} \right) \ln 2 = 5.76 \text{ J/K}$$

> It is interesting to note that the change in entropy associated with n moles of an ideal gas that doubles its volume in a free expansion is always $nR \ln 2$, regardless of the actual volume, pressure, or temperature of the gas. Deeper insight into this result is found in the statistical interpretation of entropy (Chapter 6).

It is often mistakenly thought that heat has to flow into a system for there to be an entropy change and, conversely, that any adiabatic process takes place at constant entropy, or isentropically. This example shows this not to be so. Because

$$dQ_R = TdS$$

applies only to a reversible process, a process has to be both adiabatic and reversible to be isentropic. Although free expansion is adiabatic, it is not isentropic because it is irreversible.

5.3 THE PRINCIPLE OF INCREASING ENTROPY

5.3.1 Development of the Principle

The Clausius inequality (Equation 5.1) contains the profound implication that processes can occur only if the net entropy of the universe increases or stays the same. To see how this arises, consider the cycle shown in Figure 5.5, consisting of an irreversible path i to f followed by a reversible path back to i. To be

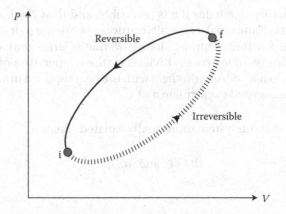

Figure 5.5 An irreversible cycle consisting of an irreversible process followed by a reversible process back to the initial state.

specific, think of a gas system. (However, the argument given here is general.) The Clausius inequality gives

$$\oint \frac{đQ}{T_0} \leq 0$$

where the equals sign applies if the path i to f is reversible and so the whole cycle is reversible.

It follows that

$$\int_i^f \frac{đQ}{T_0} + \int_{R\,f}^i \frac{đQ_R}{T} \leq 0$$

or

$$\int_i^f \frac{đQ}{T_0} \leq - \int_{R\,f}^i \frac{đQ_R}{T} = + \int_{R\,i}^f \frac{đQ_R}{T} = S_f - S_i$$

because the path f → i is reversible.

For an infinitesimal part of the process,

$$\boxed{\frac{đQ}{T_0} \leq dS}$$

5.7

where the equality sign holds if it is reversible, and then $T = T_0$. Equation 5.7 says that in an infinitesimal irreversible process between a pair of equilibrium states, there is a definite entropy change dS that is larger than the heat supplied in that irreversible process divided by the temperature of the external heat source. (Do not confuse this heat with the heat supplied in any imaginary reversible process used to calculate dS.)

Suppose now that the system is thermally isolated. Then,

$$đQ = 0 \quad \text{and} \quad dS \geq 0$$

or

$$S_f - S_i = \Delta S \geq 0 \quad (\text{thermally isolated}) \qquad 5.8$$

for a finite process. This leads to the important conclusion:

> **The entropy of a thermally isolated system increases in any irreversible process and is unaltered in any reversible process. This is the principle of increasing entropy.**
>
> **If in addition to being thermally isolated, the system is mechanically isolated from the surroundings so that no work can be done, then by the first law the internal energy U remains constant too for this condition of total isolation.**

5.3.2 Discussion and Implications

One word of warning must be given here. The principle of increasing entropy refers to *net* entropy changes. It does not say that the entropy of part of the system cannot decrease. In Figure 5.6, for example, heat flows from body A to body B at a lower temperature, both of which are contained in an adiabatic enclosure. ΔS_A is then negative but

$$\Delta S = \Delta S_A + \Delta S_B$$

is still positive, by the entropy increasing principle.

Before proceeding further, it is important to be absolutely clear as to the meaning of the entropies S_i and S_f in Equation 5.8. We have been considering an adiabatic process in which the system is changed from some initial equilibrium state with an entropy S_i to a final equilibrium state with an entropy S_f. This entropy change could be brought about by a variety of means. For example, work could be performed on the system irreversibly, or the system could consist of two parts, A and B, which are initially at different temperatures and thermally insulated from one another; heat is then allowed to flow by removing the insulator, as indicated in Figure 5.7.

If one considers only initial and final equilibrium states, the concept of an entropy change should cause no difficulty. However, consider now a system consisting of a bar inside an adiabatic enclosure, with one end initially hotter than the other, as in Figure 5.8(a). The hot end of the bar will cool, and the cold end will warm, so that the initial temperature gradient disappears, and the final state is a bar of uniform temperature. Is there an entropy change, and is Equation 5.8 still valid? The answer is yes (to both questions), but the entropy of the initial non-equilibrium state must be defined in the following way. Imagine cutting the bar into thin slices, which are then insulated from each other, as in Figure 5.8(b). The temperature of each slice may be taken as uniform over its thickness, in the limit of infinitesimal slices. Each slice may be regarded as being in an equilibrium state with a particular value of entropy determined by its mean temperature and the external pressure. (This assumes the physically reasonable assertion that the entropy of a slice does not depend on the temperature gradient, only on the mean temperature.) The entropy of

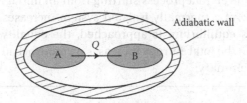

Figure 5.6   Although the entropy of a thermally isolated system can only increase or remain the same, the entropy of part of the system can decrease.

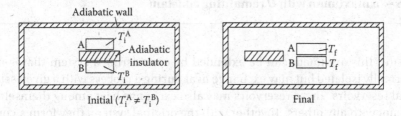

Figure 5.7   Entropy changes are calculated between equilibrium states.

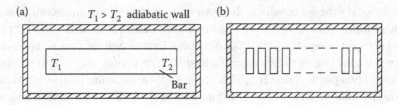

Figure 5.8 Entropy change of a bar in which there is an initial temperature gradient is calculated by dividing the bar into slices and considering the entropy change for each slice.

the whole bar may then be taken as the sum of the entropies of these slices, and this idea can be used throughout the process to consider the entropies of all the intermediate non-equilibrium states for the bar. Interested readers are referred to the book by Zemansky (1981), where the entropy change is calculated for such a bar undergoing cooling.

This concept is no different from the simple system of Figure 5.7. There, the initial entropy of the system was taken to be the sum of the entropies for the two parts, considered in isolation from each other. Then heat flowed between them until the final state of uniform temperature was reached. At any intermediate non-equilibrium state, you can still think of the entropies of the two bodies just by thermally isolating them again from each other.

We conclude that even in a process starting from an initial non-equilibrium state, the entropy of a thermally isolated system increases. It continues to increase until, as equilibrium is approached, the equality sign applies in Equation 5.8. Then the total entropy increases no more because it has reached a maximum. In summary:

> **For a system thermally isolated from the surroundings:**
> $S \rightarrow$ **a maximum**
>
> **For a system that is totally isolated from the surroundings:**
> $S \rightarrow$ **a maximum with U remaining constant**

Finally, this argument can be extended by considering a system that is not thermally isolated but may exchange heat during a process with a given set of local reservoirs. These reservoirs may also exchange heat among themselves but not with any others. Together with the original system, they form a combined system. Now surround this combined system as shown in Figure 5.9

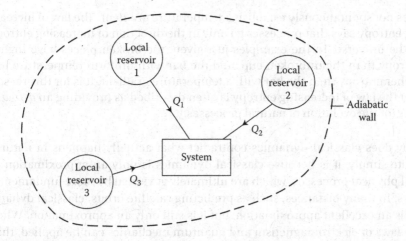

Figure 5.9   A system that exchanges heat with a set of local reservoirs only. The system and the reservoirs constitute the thermodynamic universe.

with an adiabatic wall; this will not cause any other physical changes because no heat crosses this boundary. This adiabatic enclosure contains everything that interacts during the process under consideration, and this assembly now constitutes a *thermodynamic universe* for the purpose of this argument. It should not be confused with the real universe, which may or may not be infinite and may not form an isolated system. Because this thermodynamic universe is thermally isolated,

$$\boxed{\Delta S_{universe} \geq 0}$$ 5.9

for a finite process, with the equality sign holding for a reversible process.

5.3.3 Entropy and the Arrow of Time

The entropy of a thermally isolated system can never decrease if a process provides a direction for the sequence of natural events. Newton's second law $\mathbf{F} = m\, d^2\mathbf{r}/dt^2$ is second order in time t and is unaltered by replacing t with $-t$. Thus, classical dynamics suggests that physical processes can run backwards as well as forwards, with equal likelihood. Clearly this is not so. A dropped teacup smashes into many pieces, but you have never seen the pieces spontaneously reform again into the teacup. The temperature gradient within the bar in Figure 5.8 decreases and then vanishes, but a bar with uniform temperature does not spontaneously establish a temperature gradient. The law of increasing entropy says that processes go only in the direction of increasing entropy of the universe. In the examples just given, the broken pieces have higher entropy than the unbroken cup, and the bar with uniform temperature has higher entropy than the bar with a temperature gradient. It is for this reason that the law of increasing entropy is often described as providing an *arrow of time* for the evolution of natural processes.

Why does classical dynamics contradict what actually happens in nature? Quite simply, it is because classical dynamics is only an approximation of real physical processes, which are ultimately governed by more fundamental laws. In many instances, such as predicting satellite orbits, classical dynamics is an excellent approximation, but it is still only an approximation. When the laws of electromagnetism and quantum mechanics can be applied, they supersede classical dynamics, and they are not exclusively time-reversible. For example, an accelerated charge in vacuum spontaneously radiates energy but never gains energy. An isolated atom in an excited state moves spontaneously to a lower, not higher, energy state. Statistical mechanics (Chapters 6 and 13) is essentially quantum mechanics applied to thermodynamic processes, and that is the route for understanding why the arrow of time applies in thermodynamics.

5.3.4 Example: Entropy Change for the Universe

Consider again the example (Section 5.2.2) of heating a beaker of water, with heat capacity C_P, from $T_i = 293$ K to $T_f = 373$ K. As in Section 5.3.2, the net entropy change of the universe is $\Delta S_{universe} = \Delta S_A + \Delta S_B$, where in this case A is the water and B is the reservoir at 100°C.

From Equation 5.5, the entropy change of the water is

$$\Delta S_{water} = C_P \ln\left(\frac{T_f}{T_i}\right) = C_P \ln\left(\frac{373 \text{K}}{293 \text{K}}\right)$$

Because the water's final temperature is higher than its initial temperature, the logarithm is positive and the water gains entropy.

In this process the reservoir loses an amount of heat $Q = C_P (T_f - T_i)$ irreversibly. To calculate its entropy change, imagine the reservoir losing this heat reversibly. An imaginary way of achieving this is to bring up another reservoir at a slightly lower temperature and for this heat to be transferred. Then,

$$\Delta S_{reservoir} = \int \frac{đQ_R}{T} = \frac{1}{T_f} \int đQ_R = -C_P \frac{(T_f - T_i)}{T_f}$$

$$= -C_P \left(\frac{80K}{373K} \right)$$

Notice that the temperature of the reservoir is constant at $T_f = 373$ K, and the entropy change is negative as heat flows out. Thus,

$$\Delta S_{universe} = C_P \left[\ln \left(\frac{373K}{293K} \right) - \frac{80K}{373K} \right]$$

$$= C_P (0.241 - 0.214) = 0.027 C_P$$

which is positive, as it should be for this irreversible process.

> Notice that the magnitude of the entropy increase of the water is only slightly greater than the magnitude of the entropy decrease of the reservoir. This is typical in these calculations. Despite this small difference, there is no adjustment of parameters that can lead to an overall entropy decrease. See Examples 5.1 and 5.2 at the end of this section.

It is instructive to modify this problem to ask the following question. What is $\Delta S_{universe}$ if the water is heated in two stages by placing it first on a reservoir at 50°C and, when it has reached that temperature, transferring it to a second reservoir at 100°C for the final heating?

Because the water is still being taken between the same two states, its entropy change is the same as before:

$$\Delta S_{water} = C_P \ln \left(\frac{373 \, K}{293 \, K} \right)$$

The net entropy change of the reservoirs can be found using the same method just employed:

$$\Delta S_{reservoirs} = -C_P \left(\frac{30K}{323K} + \frac{50K}{373K} \right)$$

Hence,

$$\Delta S_{\text{universe}} = C_P\left[\ln\left(\frac{373\text{K}}{293\text{K}}\right) - \frac{30\text{K}}{323\text{K}} - \frac{50\text{K}}{373\text{K}}\right] = 0.014 C_P$$

This is positive again but much less than the entropy change occurring when a single reservoir was employed. This is reasonable, because the use of two reservoirs is closer to a reversible heating, employing a number of reservoirs rather than just one.

A truly reversible cycle is accomplished using a Carnot engine. Suppose that a Carnot engine is operated between the reservoir at 100°C and the water, as in Figure 5.10. If the operating cycle of the engine is small, so that the heat đQ_2 rejected by the engine during one cycle causes only an infinitesimally small change dT in the temperature T of the water, then T does not change significantly during one cycle and the required operating conditions for a Carnot engine of operating between a pair of reservoirs exist.

Because this process is reversible, Equation 5.3 may be applied:

$$\Delta S_{\text{reservoir}} = -\int \frac{đQ_1}{T_{\text{reservoir}}} = -\int \frac{đQ_1}{373\text{K}}$$

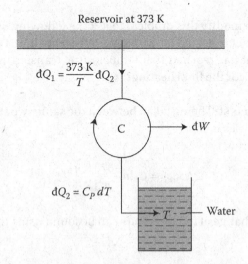

Figure 5.10   A beaker of water may be heated reversibly by operating a Carnot engine between it and a reservoir at a higher temperature.

where dQ_1 is the heat given out by the reservoir in one cycle. But

$$\frac{dQ_1}{dQ_2} = \frac{373K}{T}$$

Thus,

$$dQ_1 = dQ_2 \frac{373K}{T} = C_p dT \frac{373K}{T}$$

and hence

$$\Delta S_{reservoir} = -\int_{293}^{373} \frac{(373K)C_p dT}{(373K)T} = -C_p \ln\left(\frac{373K}{293K}\right)$$

This is the negative of the entropy change of the water. Therefore,

$\Delta S_{universe} = 0$ as it should be for a reversible process.

EXAMPLE 5.1

Consider the entropy changes that occur when water is turned into ice in your refrigerator's freezer. Assume that 100 g of water at 0°C (273 K) is turned to ice in a freezer compartment that is held at a constant –12°C (261 K). Compute the entropy reduction of the water/ice and the minimum overall entropy increase in the process.

Solution: The freezing process is accomplished by heat flowing from the water to the colder air in the freezer. The amount of heat flow is (see Section 3.3.4)

$$Q = mL = (0.10 \text{ kg})(333 \text{ kJ/kg}) = 3.33 \times 10^4 \text{ J}$$

For a constant temperature T, the entropy change is simply $\Delta S = Q/T$. Therefore the two entropy changes here are

$$\Delta S_{ice} = \frac{Q}{T} = \frac{-3.33 \times 10^4 \text{ J}}{273 \text{ K}} = -122.0 \text{ J/K}$$

$$\Delta S_{refrigerator} = \frac{Q}{T} = \frac{+3.33 \times 10^4 \text{ J}}{261 \text{ K}} = +127.6 \text{ J/K}$$

The net entropy change is 127.6 J/K – 122.0 J/K = +5.6 J/K. There is a net entropy gain, as expected.

As stated in the problem, this computed value represents the bare minimum entropy gain and is probably a low estimate. It does not take into account inefficiency in the refrigerator components and the ultimate expulsion of warm air into the kitchen.

EXAMPLE 5.2

Your bathtub contains 25 L of hot water (40°C) in a room that is at 20°C. After several hours, the bathtub water has reached room temperature. There is enough mixing over a long time that this process does not raise the room temperature appreciably. How much has the entropy of the water/room system increased?

Solution: As in Example 5.1, the first step is to compute the heat that flows from the water to the room. The specific heat capacity of water is 4186 J/(kg·°C), so

$$Q = mc\Delta T = (25 \text{ kg})(4186 \text{ J}/(\text{kg}\cdot°\text{C}))(20°\text{C}) = 2.093 \times 10^6 \text{ J}$$

With initial and final temperatures 40°C = 313 K and 20°C = 293 K, Equation 5.5 gives

$$\Delta S_{\text{water}} = C_p \ln\left(\frac{T_f}{T_i}\right) = (25 \text{ kg})(4186 \text{ J}/(\text{kg}\cdot°\text{C}))\ln\left(\frac{293 \text{ K}}{313 \text{ K}}\right) = -6910 \text{ J/K}$$

The entropy change of the air is

$$\Delta S_{\text{air}} = \frac{Q}{T} = \frac{2.093 \times 10^6 \text{ J}}{293 \text{ K}} = +7143 \text{ J/K}$$

The net entropy change is 7143 J/K – 6910 J/K = 233 J/K. Once again, there is a net entropy gain, which is inevitable in this irreversible process. As in Example 5.1, this entropy change includes only a small part of the process. For example, it ignores the (enormous!) entropy generated by the water heater that raised the temperature of the water to 40°C in the first place.

Comparing the results of Examples 5.1 and 5.2, it is interesting to note that a similar outcome resulted from the first case, a process generated by a human-built machine, and the second case, a slow spontaneous natural process. In a sense, refrigerators and freezers are designed to decrease entropy in a small region of space by lowering the temperature there. However, this can only come at the expense of increasing entropy in another space. The numerical values computed in the two examples are also consistent with the example of heating water done at the start of Section 5.3.4. That is, the new entropy increase in the universe is fairly small compared with the individual entropy changes (increase and decrease) of the two bodies involved. Nevertheless, the overall increase is inevitable.

5.4 ENTROPY–TEMPERATURE DIAGRAMS

The thermodynamic state of a system can be specified by any pair of independent state functions. In particular, a state is equally well specified by the pair S and T as by the pair P and V. Just as it is possible to represent a reversible process as a line joining a succession of equilibrium states on a PV diagram, the same can be done on a TS diagram. However, the form of the line is very simple for certain useful processes.

Equation 5.4 shows that a reversible adiabatic process is an isentropic one, and therefore such a process is represented on a TS diagram as a straight line parallel to the T axis, as shown in Figure 5.11. A reversible isothermal process is represented by a straight line parallel to the S axis. Thus, the cycle for a

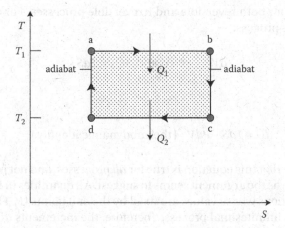

Figure 5.11   *TS* diagram for a Carnot cycle.

Carnot cycle is a rectangle on a TS plot. Compare this with Figure 4.1, the PV diagram for the Carnot cycle.

As for any reversible process,

$$Q = \int_R T dS$$

Therefore, the net heat absorbed in a Carnot cycle is given by the area shaded in the figure. This fact helps make TS plots of enormous value in engineering. You can think of this as analogous to the fact that the net work done in a cycle is the area enclosed on a PV diagram. In both cases the sign (of the net heat or net work) depends on the direction of the path taken around the cycle. (See Problem 5.17.)

5.5 THE THERMODYNAMIC IDENTITY

The first and second laws of thermodynamics can be combined to obtain an important equation in thermodynamics.

The differential form of the first law is

$$dU = đQ + đW$$

which is true for both reversible and irreversible processes. For an infinitesimal reversible process,

$$đW = -PdV \text{ and } đQ_R = TdS$$

Thus,

$$dU = TdS - PdV \quad \text{(thermodynamic identity)} \qquad 5.10$$

We now argue that this equation is true for *all* processes, and not just for reversible processes, as the argument seems to suggest. All quantities in Equation 5.10 are state functions whose values are fixed by the end points (P, T) and $(P + dP, T + dT)$ of the infinitesimal process. Therefore, the increments dU, dS, and dV are fixed and do not depend on the path joining the end points. As a result, any relation between them is independent of whether or not the process is reversible.

This is a significant advance, because Equation 5.10 is a general relation among P, V, T, and S that holds for all paths between a pair of infinitesimally close equilibrium states, whether or not they are reversible. The relation expressed in Equation 5.10 is called the *thermodynamic identity*. It is sometimes called the *central equation of thermodynamics*, to stress its importance. This is no exaggeration. The whole of the science of thermodynamics is dependent on this equation, just as the whole of mechanics is dependent on Newton's laws. Because it is an identity, and you do not have to ask whether the process we are considering is reversible or irreversible, it follows that the equations derived from it are generally true.

Later in this book you will see modifications of the thermodynamic identity that are useful in other situations. These arise because it is often useful to consider forms of energy other than its internal energy U. In Chapter 7 the thermodynamic identity is recast (separately) in terms of enthalpy H, Gibbs free energy G, and Helmholtz free energy F. In Chapter 11 the effects of open systems are considered, and there the thermodynamic identity is adjusted to include the chemical potential function μ.

Finally, note that Equation 5.10 considers only volume work PdV. If there are other kinds of work, these must be included in the thermodynamic identity. For example, if there were also magnetic work, the equation would have to be modified to

$$dU = TdS - PdV + B_0 d\mathcal{M}$$

5.6 OTHER EXAMPLES OF ENTROPY CALCULATIONS

Entropy is widely useful in thermodynamics. Before going on we present two other examples of how entropy can be computed in real physical systems.

5.6.1 Entropy of an Ideal Gas

Although the demonstration of the power of Equation 5.10 has to wait until Chapter 7, we can give an example here of its use to determine an expression for the entropy of an ideal gas in terms of the volume and temperature.

For an ideal gas, where $U = U(T)$ only,

$$C_V = \left(\frac{\partial U}{\partial T}\right)_V = \frac{dU}{dT}$$

and so Equation 5.10 becomes

$$TdS = C_V dT + PdV$$

Using the equation of state $PV = nRT$,

$$TdS = C_V dT + \frac{nRT}{V} dV$$

The problem is best considered using molar quantities for the extensive parameters and representing them with lower-case variables, s (for entropy), c_v (for molar specific heat), and $v = V/n$ for molar volume:

$$ds = \frac{c_v}{T} dT + \frac{R}{v} dv$$

Integrating, the molar entropy of an ideal gas is

$$s = c_v \ln T + R \ln v + s_0 \qquad \text{5.11}$$

where s_0 is an integration constant that disappears when entropy differences are taken.

5.6.2 Entropy of a Black Hole

A black hole is an astronomical object that has its mass concentrated in such a small region of space that its gravitational field prevents light (or other electromagnetic radiation) from escaping. Black holes are found throughout the universe, for example, as remnants of supernova events and large ones at the cores of many galaxies.

Generally, the entropy of a system tends to increase with the system's mass. (Chapter 6 will better explain this fact from a statistical perspective.) The ideal gas entropy in Equation 5.11 is, for example, consistent with this principle, as the entropy grows with the number of moles n and the system's heat capacity C_V. A black hole cannot be understood classically, but using quantum mechanics and general relativity, Stephen Hawking found that a black hole of mass M has entropy

$$S = \frac{8\pi^2 k_B GM^2}{hc} \qquad 5.12$$

where G is the universal gravitation constant, k_B the Boltzmann constant, c the speed of light, and h Planck's constant. Notice that the entropy increases as the square of the mass.

In Equation 5.12, we have not bothered to include the usual constant term s_0. This is justified by the unusually large value for any possible black hole entropy. For example, for a black hole with a mass of 10 solar masses (or 2.0×10^{31} kg), Equation 5.12 gives an entropy of 1.5×10^{56} J/K. It is difficult to put this number into perspective, but it is immensely larger than the entropy you might encounter in everyday experience or laboratory work, for example an ideal gas with entropy given by Equation 5.11.

Problems

5.1 A bucket containing 5.0 kg of water at 25°C is put outside a house so that it cools to the temperature of the outside air at 5°C. What is the entropy change of the water? [c_p for water = 4.19 kJ/(kg· °C).]

5.2 Five kg of water at 25°C is added to 10.0 kg of water at 85°C. After the mixture has reached equilibrium, how much has entropy changed? (Assume no energy is exchanged between the water and its surroundings.)

5.3 Two systems that have the same heat capacity C_V but different initial temperatures T_1 and T_2 (with $T_2 > T_1$) are placed in thermal contact with each other for a brief time, so that some heat flows but the temperature of neither system changes appreciably. Show that there is a positive net entropy change associated with this heat flow.

5.4 Calculate the entropy change for each of the following: (a) 10 g of steam at 100°C and a pressure of 1 atm condensing into water at the same temperature and pressure. (The latent heat of vaporization of water at that temperature is 2260 J/g.); (b) 10 g of water at 100°C and a pressure of 1 atm cooling to 0°C at the same pressure. (The average specific heat of water between 0°C and 100°C is 4.19 J/g.); and (c) 10 g of water at 0°C and a pressure of 1 atm freezing into ice at the same pressure and temperature. (The latent heat of fusion of ice is 333 J/g.)

5.5   The low-temperature molar specific heat of diamond varies with temperature as

$$c_V = 1.94 \times 10^3 \left[\frac{T}{\theta}\right]^3 \text{ J/(mol·K)}$$

where the Debye temperature $\theta = 1860$ K. What is the entropy change of 1.0 g of diamond when it is heated at constant volume from 4–300 K? (The atomic mass of carbon is 12.0 g/mol.)

5.6   An electric current of 10 A flows for 1 min through a resistor of 20 Ω which is kept at 10°C by being immersed in running water. What is the entropy change of the resistor, the water, and the universe?

5.7   A thermally insulated resistor of 20 Ω has a current of 2.5 A passed through it for 1.5 s. It is initially at 20°C. The resistor's mass is 5.0 g, and c_p for the resistor is 0.80 kJ/(kg· °C). (a) What is the final temperature? (b) What is the entropy change of the resistor and the universe? (Hint: In the actual process, dissipative work is done on the resistor. Imagine a reversible process taking it between the same equilibrium states.)

5.8   An ideal gas has a molar specific heat given by $c_V = A + BT$ where A and B are constants. Show that the change in entropy per mole in going from the state (V_1, T_1) to the state (V_2, T_2) is

$$\Delta S = A\ln\left(\frac{T_2}{T_1}\right) + B(T_2 - T_1) + R\ln\left(\frac{V_2}{V_1}\right).$$

5.9   A 50-kg bag of sand at 25°C falls 10 m onto the pavement and comes to an abrupt stop. What is the entropy increase of the sand? Neglect any transfer of heat between the sand and the surroundings and assume that the thermal capacity of the sand is so large that its temperature is unchanged. (Hint: Consider the following: (i) What is the dissipative work done on the sand? (ii) What is the change in the internal energy of the sand? (iii) What is the entropy change associated with this ΔU at constant T? The sand does no external work as it deforms when it hits the pavement; only its shape changes, not its volume.)

5.10 Two moles of an ideal gas undergo a free expansion, tripling the volume. What is the entropy change of (a) the gas and (b) the universe?

5.11 Two equal quantities of water of mass m and at temperatures T_1 and T_2 are adiabatically mixed together, the pressure remaining constant. Show that the entropy change of the universe is

$$\Delta S = 2mc_p \ln\left(\frac{T_1 + T_2}{2\sqrt{T_1 T_2}}\right)$$

where c_p is the specific heat of water at constant pressure. Show that $\Delta S \geq 0$. (Hint: $(a - b)^2 \geq 0$ for any real a and b.)

5.12 Consider two identical bodies of heat capacity C_p and with negligible thermal expansion coefficients. Show that when they are placed in thermal contact in an adiabatic enclosure their final temperature is $(T_1 + T_2)/2$ where T_1 and T_2 are their initial temperatures. Now consider these two bodies being brought to thermal equilibrium by a Carnot engine operating between them. The size of the cycle is small, so that the temperatures of the bodies do not change appreciably during one cycle; thus the bodies behave as reservoirs during one cycle. Show that the final temperature is $(T_1 T_2)^{1/2}$. (Hint: What is the entropy change of the universe for this second process?)

5.13 A semipermeable membrane is one that allows the passage of one type of molecule. At equilibrium the gas pressures on either side of such a membrane are equal. Such membranes exist.

Consider a mixture of two ideal gases A and B contained in the left half of the box as shown in Figure 5.12(a). There is a vacuum in the

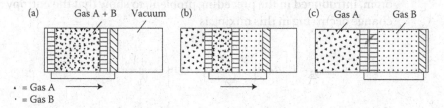

(a)   Gas A + B   Vacuum   (b)   (c)   Gas A   Gas B

$\cdot$ = Gas A
$\cdot$ = Gas B

Figure 5.12

right half. The box is fitted with a pair of coupled sliding pistons; the left one is permeable to A only, while the right one is impermeable to both. The box is divided into two with a partition permeable to B only. Now slide the coupled pistons slowly to the right as shown in Figure 5.12(b) so that, eventually, the two gases separate reversibly. They will finally each occupy a volume equal to the original volume of the mixture. This is shown in Figure 5.12(c). Let this process occur isothermally. (a) By considering the pressures due to each gas on either side of the membranes, show that the net force on the coupled pistons is zero. (b) The heat flowing into the system in this isothermal reversible process is $Q = T(S_f - S_i)$ where S_i and S_f are the initial and final entropies. By now applying the first law, show that $S_i = S_f$.

This result is known as Gibbs's theorem. It says:

In a mixture of ideal gases, the entropy is the sum of the entropies that each gas would have if it alone occupied the whole volume.

In other words,

$$S_{A+B}(T,V) = S_A(T,V) + S_B(T,V)$$

5.14   From Equation 5.11, for n moles of an ideal gas the entropy is

$$S = nc_v \ln T + nR\ln\left(\frac{V}{n}\right) + S_0 \qquad\qquad 5.13$$

Now n_A moles of an ideal gas A of volume V_A and temperature T are separated from n_B moles of another ideal gas B of volume V_B at the same temperature T (see Figure 5.13(a)). The partition is removed so that the gases mix isothermally at the temperature T, the mixture then occupying the volume $V_A + V_B$ (see Figure 5.13(b)). (a) Use Gibbs's theorem, introduced in the preceding problem, to show that the entropy change occurring in this mixing is

$$\Delta S_{\text{mixing}} = R\left[n_A \ln\left(\frac{V_A + V_B}{V_A}\right) + n_B \ln\left(\frac{V_A + V_B}{V_B}\right)\right]$$

(b) Suppose that the gases are identical. Clearly, on removing the partition, there can now be no entropy change, because the physical

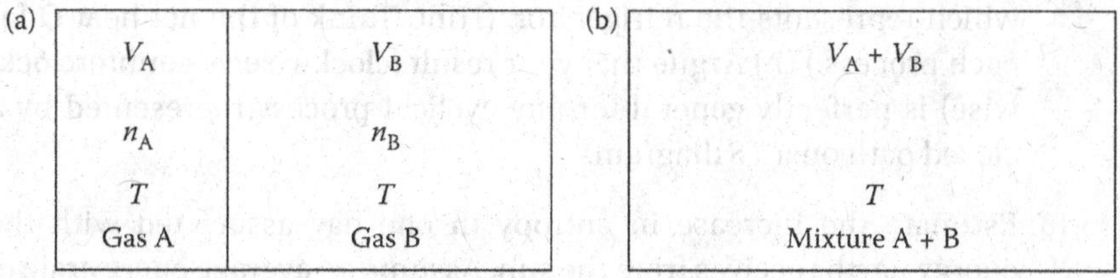

(a)

V_A	V_B
n_A	n_B
T	T
Gas A	Gas B

(b)

$V_A + V_B$
T
Mixture A + B

Figure 5.13

system is unchanged. However, the result just proved in (a) gives $\Delta S_{\text{mixing}} \neq 0$! This is known as the *Gibbs paradox*. Is the result given in (a) valid for identical gases and if not, why not? (Hint: Consider how Gibbs's theorem was proved.) (c) Obtain the correct expression

$$\Delta S_{\text{mixing}} = \left(n_A + n_B\right)R\ln\left(\frac{V_A + V_B}{n_A + n_B}\right) - n_A R\ln\left(\frac{V_A}{n_A}\right) - n_B R\ln\left(\frac{V_B}{n_B}\right)$$

for the entropy of mixing of identical gases by applying Equation 5.11 to the three volumes V_A, V_B, and $V_A + V_B$, all containing the same gas. (d) By using the fact that, for identical gases,

$$\frac{V_A + V_B}{n_A + n_B} = \frac{V_A}{n_A} = \frac{V_B}{n_B}$$

show that the entropy of mixing given in (c) is indeed zero. (The Gibbs paradox is discussed in the book by Chambadal (1973).)

5.15 One mole of helium gas is initially at $P_0 = 1.0$ atm and $T_0 = 273$ K. (a) Compute the entropy change if the gas is heated at constant pressure to temperature 400 K. (b) Starting again from the initial state (P_0, T_0), what is the entropy change if the gas expands isothermally to twice its original volume?

5.16 A refrigerator with coefficient of performance 3.5 uses 2.0 kWh of electrical energy to keep the refrigerator compartment at 4°C while expelling heat to a kitchen at 20°C. How much entropy is generated in one day?

5.17 A Carnot cycle like the one in Figure 5.11 might represent, for example, a Carnot heat engine or a Carnot refrigerator. The difference between the two is the direction of the path taken. (a) Explain which path (clockwise or counterclockwise) represents the heat engine and which represents the refrigerator. (Hint: Think of the net heat Q for each process.) (b) Argue that your result (clockwise vs counterclockwise) is perfectly general for any cyclical process represented by a closed path on a TS diagram.

5.18   Estimate the increase in entropy in one day associated with the energy Earth receives from the sun. Assume an average energy rate of 240 W/m² on the surface of Earth facing the sun and an average surface temperature of 15°C.

REFERENCES

Chambadal, P., *Paradoxes of Physics*, Transworld, London, 1973.

Zemansky, M.W. and Dittman, R.H., *Heat and Thermodynamics*, McGraw Hill, New York, 1981.

Chapter 6: Statistical Mechanics

The study of thermodynamics is advanced substantially by considering how macroscopic thermal behavior depends on the behavior of atoms and molecules. It is impossible to measure and predict the motions of a large number of individual molecules, on the order of Avogadro's number or more. For that reason it is necessary to apply the laws of statistics. This way of studying thermal physics is called *statistical thermodynamics* or *statistical mechanics*.

6.1 INTRODUCTION TO PROBABILITY AND STATISTICS

Fortunately, the statistical properties of physical systems are based largely on some simple rules of probability, which are well understood. Our approach will be to introduce some important rules and concepts from probability for smaller systems and then generalize to larger ones.

6.1.1 Probability in a Two-State System

Tossing a coin is a familiar and simple example that illustrates some basic concepts in probability. Suppose you have a fair coin that has an equal chance or showing heads (H) or tails (T) when tossed. If you toss the coin twice in succession and record the result each time, there are four possible outcomes: HH, HT, TH, and TT (where for example HT means H on the first toss and T on the second).

In the experiment just described, what is the probability that you will observe a total of 2, 1, or 0 heads in the two tosses? Notice that out of the four results there is one way to get two heads (HH), one way to get zero heads (TT), but there are two ways (HT and TH) to get one head. A general rule is that the probability of a particular result such as 2, 1, or 0 heads is the number of ways

DOI: 10.1201/9781003299479-6

to obtain that result divided by the total number of outcomes, in this case four. (This is based on the idea that on any particular pair of tosses each of the four outcomes is equally likely.) Therefore the probabilities are

$$P(2) = \frac{1}{4} = 0.25 \quad P(1) = \frac{2}{4} = 0.50 \quad P(0) = \frac{1}{4} = 0.25$$

where for example the notation $P(2)$ means the probability of two heads. Notice that, by convention, the sum of the probabilities is arranged to be exactly 1, which we will take to be a general rule.

> **Sometimes you see probabilities expressed as a percentage, such as 25% for probability 0.25. Because we have defined probabilities as fractions, we will represent them only as fractions or in the equivalent decimal form but not as percentages, so as to avoid confusion.**

6.1.2 Ideal Gases and Multiplicity

The example of tossing coins is analogous to the two-molecule gas in the two-sided chamber shown in Figure 6.1(a), where the molecules are *distinguishable*. Here, and for the remaining part of this section (but not Section 6.2), we consider each space within the chamber to be large compared with the size of a molecule (or its wavefunction). Thus, the gas molecules can be considered distinguishable and assigned particular labels.

The two molecules are free to move throughout a rectangular box. For a classical gas, it is assumed that collisions with the walls and with the other molecule

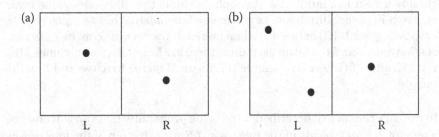

Figure 6.1 (a) A two-molecule gas in a two-sided chamber. (b) A three-molecule gas in the same chamber. Note that the vertical center line between the two sides of this chamber simply denotes the two sides of the chamber and is not a physical barrier.

are elastic. With these assumptions, the probability of a given molecule being in either the left half (L) or right half (R) is exactly one half. At random times you might observe both molecules to see how many are in each half. By analogy with the coin toss, a single measurement of the number of molecules on the left half will yield 2, 1, or 0 with probabilities 0.25, 0.50, and 0.25, respectively.

What happens when the number of molecules is increased? Figure 6.1(b) shows a three-molecule gas. A measurement of the positions (L or R) of the three molecules has the following eight outcomes: LLL, LLR, LRL, RLL, LRR, RLR, RRL, and RRR. Now there is one way (LLL) to find three molecules on the left side, but there are three ways (LLR, LRL, and RLL) to find two molecules on the left side, and so on. The probability $P(n)$ of finding n molecules on the left is given by

$$P(3)=\frac{1}{8}=0.125 \quad P(2)=\frac{3}{8}=0.375 \quad P(1)=\frac{3}{8}=0.375 \quad P(0)=\frac{1}{8}=0.125$$

Before generalizing the results, some definitions are in order. A particular ordering of the three molecules (such as LLR) is called a *microstate*, and the set of orderings that yield the same result (such as two on the left) is called a *macrostate*. The number of microstates in any macrostate is called the *multiplicity* and designated by the symbol Ω. For example, in the three-molecule gas the macrostate with two molecules on the left has multiplicity $\Omega = 3$.

In general, the probability of any macrostate is given by the multiplicity of that macrostate divided by the sum of all the multiplicities. Symbolically,

$$P(n)=\frac{\Omega(n)}{\sum_i \Omega(i)} \qquad\qquad 6.1$$

where the sum i is carried out over all macrostates. For the three-molecule gas, the four probabilities computed above follow from this general rule. It is straightforward to show that the sum of all the probabilities is one, as required.

The three-molecule gas is still not very useful, but these computations are easily generalized to a gas of N molecules. The multiplicity $\Omega(N, n)$ of the macrostate with n out of the N molecules on one side (left or right) is

$$\Omega(N,n)=\frac{N!}{n!(N-n)!} \qquad\qquad 6.2$$

This result, familiar in probability theory, is sometimes read "N choose n" because you are choosing n out of the total N to be in one particular place. It is expressed in the shorthand notation

$$\Omega(N,n) = \binom{N}{n}$$

The results for both the two- and three-molecule gases are consistent with Equations 6.1 and 6.2.

The results $\Omega(N, n)$ are the same as the binomial coefficients from algebra. For example, $(x + y)^3 = x^3 + 3x^2 y + 3xy^2 + y^3$, with the coefficients 1, 3, 3, 1 matching $\Omega(N, n)$ with $N = 3$. You may have seen the coefficients displayed graphically in *Pascal's triangle*.

6.1.3 Larger Systems

It is useful to apply the results in the preceding section to an N-molecule gas, where in most situations N becomes a very large number. The multiplicity of the state with n molecules in the left half of the box is given by Equation 6.2. As N grows, the multiplicity becomes more sharply peaked around its maximum value at $n = N/2$. Table 6.1 illustrates this for some larger values of N. In each case the peak value $\Omega(N, N/2)$ is compared with $\Omega(N, n)$ just $0.1N$ away from the peak. These results clearly show the trend for increasing N. The same trend is shown graphically in Figure 6.2. As the number of particles in the sample grows, the multiplicity function grows sharper.

Now consider what happens if the gas is the air in a typical room, with N perhaps on the order of 10^{27} for a room full of air under normal conditions.

TABLE 6.1   Multiplicities of Selected States for an N-molecule Monatomic Gas

N	$\Omega(N, N/2)$	$\Omega(N, N/2 \pm 0.10N)$	Ratio
10	252	210	0.83
100	1.0×10^{29}	1.4×10^{28}	0.14
1000	2.7×10^{299}	5.0×10^{290}	2×10^{-9}

Note: The ratio in the final column is the ratio of the two preceding columns, showing the relative likelihood of a 10% deviation from the peak value.

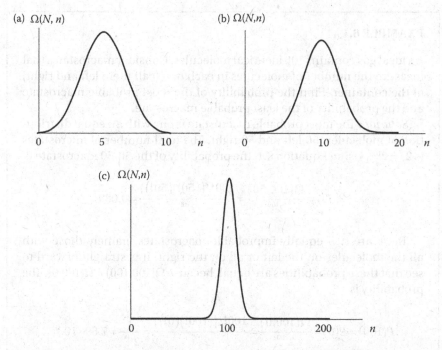

Figure 6.2 Multiplicity function $\Omega(N, n)$ for (a) $N = 10$; (b) $N = 20$; and (c) $N = 200$ particles. Even with a fairly small number of particles, the sharpening of the multiplicity function with increasing N is evident.

The multiplicities are too large to present in a table, but the trend you have seen illustrates that it is very very unlikely that there will *ever* be any significant deviation from $n = N/2$. There are certainly fluctuations in n as the gas molecules zip around, but those deviations are too small relative to $N \approx 10^{27}$ for you to notice or to measure as a local pressure deviation. Thankfully, after reading this section you can go on living with every expectation that the air you need to breathe will not spontaneously flood the other side of the room!

Conversely, the same statistical argument explains the process of free expansion (see Sections 3.5.1 and 5.2.3). A gas initially confined to volume V and then suddenly allowed to expand into volume $2V$ will do so readily, because it is much more likely that the molecules will be distributed throughout the new allowed volume $2V$ than the smaller volume V. Free expansion will be considered again from a statistical viewpoint in Section 6.2.

EXAMPLE 6.1

An ideal gas contains 100 identical molecules. Consider macrostates that measure the number of molecules in each half (call them left and right) of the container. Find the probability of the most probable macrostate and the probability of the least probable macrostate.

Solution: The most probable macrostate is one with an equal distribution of molecules, 50 left and 50 right. The total number of microstates is $2^N = 2^{100}$. Using Equation 6.1, the probability of the 50/50 macrostate is

$$P(50) = \frac{\Omega(100,50)}{2^{100}} = \frac{100!/\big((50!)(50!)\big)}{2^{100}} = 0.080$$

There are two equally improbable macrostates, namely those with all the molecules on the left or all on the right. It is straightforward to see that these probabilities are equal, because $\Omega(100,100) = \Omega(100,0)$. The probability is

$$P(100) = P(0) = \frac{\Omega(100,0)}{2^{100}} = \frac{100!/\big((100!)(0!)\big)}{2^{100}} = \frac{1}{2^{100}} = 7.89 \times 10^{-31}$$

These numerical results show a probability distribution that is sharply peaked around the maximum value, as illustrated in Figure 6.2.

6.2 MICROSCOPIC VIEW OF ENTROPY

In Section 6.1 we considered the statistical distribution of an N-molecule gas based on measurements of which half of the container each molecule occupies. This is only a crude start and does not begin to provide a complete statistical description of the gas's thermal properties.

6.2.1 Phase Space

To begin understanding the gas's thermal properties, note that the microscopic description of a gas involves each molecule's position, for example (x, y, z) in Cartesian coordinates, and its momentum components (p_x, p_y, p_z).

Together these six variables constitute a six-dimensional *state space* or *phase space*. For an *N*-molecule gas, the appropriate phase space has 6*N* dimensions.

On the scale of single molecules, it is necessary to use quantum mechanics rather than classical mechanics. Heisenberg's uncertainty principle says that for a particle in one-dimensional motion, position x and momentum p_x cannot be simultaneously measured exactly but rather have uncertainties Δx and Δp_x, with the restriction that the product $\Delta x \Delta p_x$ has a minimum value on the order of Planck's constant h. For three-dimensional motion, the same relation holds for each pair of position–momentum components, so the product $\Delta x \Delta y \Delta z \Delta p_x \Delta p_y \Delta p_z$ has a minimum value on the order of h^3.

> **Note: We say on the order of h^3 because the exact minimum value depends on the type of wave function being used to describe the particle. For the purposes of this discussion, there is no need to be more precise.**

The uncertainty in measurement is illustrated graphically in Figure 6.3. It is impossible to visualize a six-dimensional space in three-dimensional diagrams, so for simplicity just two spatial dimensions and two velocity

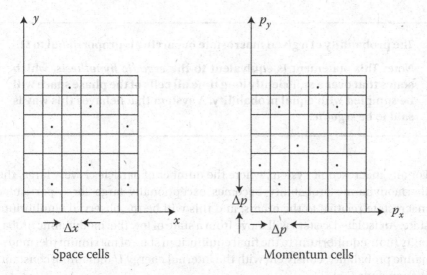

Figure 6.3 A two-dimensional schematic representation of phase space. A distribution of all the particles over the cells gives rise to a definite macroscopic state. In three dimensions the space cells are of volume Δx^3 and the momentum cells of volume Δp^3.

components are represented on separate diagrams. In the diagram on the left, the uncertainties Δx and Δy divide space into cells. Each dot represents a particle that has its position (x, y) within that cell. Similarly, the diagram on the right shows momentum space divided into cells, with each dot representing a particle with momentum (p_x, p_y).

The real three-dimensional space, with its six-dimensional "volume" (called a *hypervolume*), is then split into cells having hypervolume $(\Delta x)^3 (\Delta p)^3$, where, by symmetry the three spatial dimensions are assumed equivalent (Δx) and the three momentum dimensions are also equivalent (Δp). The statistical description of the system is based on counting the number of particles that fall into each hypervolume cell.

6.2.2 Statistical Entropy

In terms of the kinds of states described in Section 6.1, a microstate of a system is defined by having a certain number of particles in each phase space cell. A macrostate, on the other hand, has a particular set of state variables: temperature, entropy, and so on. In general there are many different microstates that give rise to the same macrostate, and this is the multiplicity Ω of the macrostate. This leads to the key statement of statistical mechanics:

The probability of a given macrostate occurring is proportional to Ω.

Note: This statement is equivalent to the *ergodic hypothesis*, which states that over a sufficiently long time all cells of the phase space will be sampled with equal probability. A system that behaves this way is said to be *ergodic*.

For any macroscopic system, where the number of particles is very large, the thermodynamic probability becomes exceptionally large for a particular macrostate relative to the others, and this will be the observed equilibrium state. An isolated system will move from a state of low thermodynamic probability (non-equilibrium) to the final equilibrium state of maximum thermodynamic probability, consistent with the internal energy U remaining constant. We conclude that

$$\Omega \rightarrow \text{a maximum}$$

This is a clue to the meaning of entropy. Remember from Chapter 5 that, for an isolated system,

$$S \to \text{a maximum}$$

while U remains constant. Additionally, S is an extensive quantity, so that the entropy of two separate systems is $S_1 + S_2$. If the number of ways of realizing the first system is Ω_1 and Ω_2 for the second, then the number of ways of realizing both systems together is

$$\Omega = \Omega_1 \Omega_2$$

This leads to the definition of statistical entropy:

$$S = k_B \ln \Omega \qquad\qquad 6.3$$

where k_B is the Boltzmann constant. The factor $\ln \Omega$ is required, so that $\ln (\Omega_1 \Omega_2) = \ln \Omega_1 + \ln \Omega_2$, to satisfy the additive property of entropy. It will be clear later why ln is chosen over log. The factor k_B is not as obvious, but it is correct on dimensional grounds and will be justified by the correct results to which it leads in thermodynamics. Equation 6.3 is famous and is known as the *Boltzmann relation*. To summarize this important result:

> **The microscopic viewpoint interprets the increase of entropy for an isolated system as a consequence of the natural tendency of the system to move from a less probable to a more probable state.**

Remarkably, the statistical version of entropy is equivalent to the thermodynamic entropy defined in Chapter 5. We will not prove this fact, but rather it will be illustrated in examples that follow in Sections 6.2.4 and 6.2.5.

> **A version of the definition of statistical entropy (equivalent to Equation 6.3) is carved on Boltzmann's tombstone in Vienna.**

6.2.3 Entropy and Disorder

Sometimes Ω is identified as a measure of "disorder" in the system. This implies that you should expect the disorder of an isolated system to increase to the maximum amount permissible, which occurs when equilibrium is reached. To see what this means, consider the microstate in which all the particles are in one cell in phase space. This is a highly ordered arrangement in phase space, which can be achieved in only one way with $\Omega = 1$ and $S = 0$. It is a highly ordered arrangement in real space, too, with all the particles in the same place and moving with identical velocities. The particles will spread out from this highly ordered state, occupying more cells in phase space and lessening the order or increasing the disorder in that space. The multiplicity will increase from 1 to a large value, with the entropy increasing accordingly. It is in this sense that Ω is a measure of disorder.

However, you should note this word of caution. Disorder is a subjective property, while multiplicity and entropy are exact physical quantities. Apart from simple cases like the one described above, the connection between entropy and disorder is not always clear, and you can fool yourself by trying to make predictions based on a subjective assessment of order and disorder rather than quantitative laws.

With this in mind, it is useful to examine the agreement between the macroscopic and microscopic viewpoints in two specific examples.

6.2.4 Entropy Change in Free Expansion of an Ideal Gas: Microscopic Approach

You know that there is no temperature change in a free expansion of an ideal gas, and so the mean kinetic energy and root-mean-square momentum $p_{rms} \equiv \left(\overline{p^2}\right)^{1/2}$ of the molecules remains unchanged. Consider such an expansion in which the volume is doubled.

The momentum part of phase space is a hypercube of volume on the order of p_{rms}^3. Therefore, the number of momentum cells that can be occupied is $\approx p_{rms}^3/\Delta p^3$, and this number does not change upon expansion, because p_{rms} is unchanged. However, the number of *space* cells that can be occupied *doubles* from $V/\Delta x^3$ to $2V/\Delta x^3$, where V is the original volume. This means that if the number of possible arrangements for fitting the molecules in the cells before the expansion is Ω, after expansion it is now greater than this. In fact, for N distinguishable molecules distributed among q distinct cells $\Omega = q^N$, so after this expansion it becomes $2^N \Omega$. Thus

$$\Delta S = k_B \ln(2^N \Omega) - k_B \ln \Omega = k_B \ln 2^N$$

Using the properties of logarithms:

$$\Delta S = N k_B \ln 2 = n R \ln 2 \qquad\qquad 6.4$$

with the result expressed both in terms of N molecules and n moles. This matches the result obtained earlier (Equation 5.6), so in this case the statistical and thermodynamic approaches yield identical results.

6.2.5 Entropy of an Ideal Gas: Microscopic Approach

In Chapter 5 you saw that the thermodynamic identity can be used to obtain an expression for the entropy of an ideal gas (Equation 5.11). The same result can be obtained from microscopic considerations. For simplicity, consider a monatomic gas, so that the atoms have only translational degrees of freedom.

The atoms of the gas have to be placed into the cells of phase space subject to the following two restrictions:

1. All the atoms have to be contained in a box of volume V.

2. The total energy of the atoms of mass m is fixed at U and is all kinetic, with

$$U = \sum_i \frac{p_i^2}{2m}$$

where the summation is over the N atoms.

The total number of ways Ω of filling up the cells in phase space is the product of the number of ways Ω_{space} the space cells[1] of volume Δx^3 can be filled multiplied by the number of ways $\Omega_{momentum}$ the different momenta cells of volume Δp^3 can be filled. Thus

$$\Omega = \Omega_{space}\Omega_{momentum}$$

First, think about how to calculate Ω_{space}. As a visual guide, Figure 6.4 shows just two dimensions of space cells, with $\Delta x = \Delta y$ for convenience. In three dimensions, each atom has $V/\Delta x^3$ distinct locations in the box. Thus

---

[1] The use of the word cell in statistical mechanics should be strictly confined to an elementary volume $\Delta x^3 \Delta p^3$ of phase space.

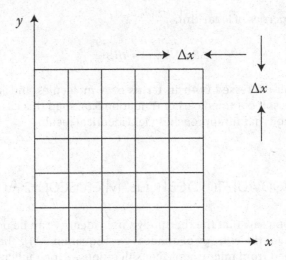

Figure 6.4   Space cells of volume Δx^2 in two dimensions, which translates to Δx^3 in three dimensions.

$$\Omega_{\text{space}} = \left[\frac{V}{\Delta x^3} \right]^N$$

Next, consider how to calculate Ω_{momentum}. Although each atom is not confined to a finite "momentum box," the atoms have a root-mean-square momentum p_{rms}. Because $U = N\bar{E}$, the relationship between U and p_{rms} is

$$U = N \frac{p_{\text{rms}}^2}{2m}$$

For the purpose of this calculation, we may take the atoms as being confined within a momentum box of side p_{rms} as shown in Figure 6.5. The number of cells in the momentum box is $(p_{\text{rms}}/\Delta p)^3$ for each atom. Thus

$$\Omega_{\text{momentum}} \approx \left(\frac{p_{\text{rms}}}{\Delta p} \right)^{3N}$$

Multiplying these two results,

$$\Omega = \Omega_{\text{space}} \cdot \Omega_{\text{momentum}} \approx \left[\frac{p_{\text{rms}}^3 V}{\Delta x^3 \Delta p^3} \right]^N$$

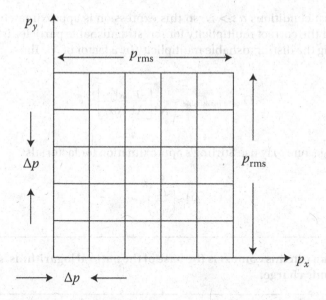

Figure 6.5 Momentum cells of volume Δp^2 in two dimensions, which translates into Δp^3 in three dimensions.

However, this argument has overcounted the ways of filling the phase space cells, because it has assumed that the atoms are *distinguishable*, just as if they are labeled with a number. The two situations depicted in Figure 6.6 are clearly the same physically. To find the correct multiplicity for indistinguishable particles, notice that there are fewer ways of arranging the N identical atoms in a given set of distinct boxes. For N indistinguishable particles distributed among q distinct cells,

$$\Omega_{\text{indistinguishable}} = \frac{(q+N-1)!}{(N!)(q-1)!}$$

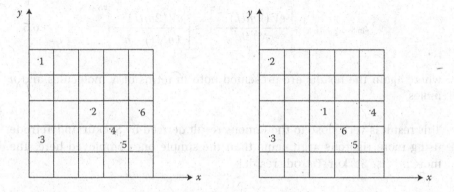

Figure 6.6 An illustration of two equivalent arrangements in phase space.

Under most conditions $q \gg N$, so this expression is approximately equal to $q^N/N!$, and the correct multiplicity for indistinguishable particles is obtained by dividing the distinguishable multiplicity by a factor of $N!$. Thus

$$\Omega_{\text{indistinguishable}} \approx \frac{1}{N!} \left[\frac{p_{\text{rms}}^3 V}{\Delta x^3 \Delta p^3} \right]^N$$

In N is large, one may use Stirling's approximation for factorials:

$$N! \approx \left(\frac{N}{e} \right)^N$$

> Note that e in this context is the base of the natural logarithms, not the electronic charge.

Therefore

$$\Omega_{\text{indistinguishable}} \approx \left[\frac{eVp_{\text{rms}}^3}{N(\Delta x \Delta p)^3} \right]^N$$

$$\approx \left[\frac{eV(2mU)^{3/2}}{N^{5/2}(\Delta x \Delta p)^3} \right]^N \quad \text{using} \quad p_{\text{rms}} = \left(\frac{2mU}{N} \right)^{1/2}$$

Using the Heisenberg principle with $\Delta x \Delta p = h$ for minimum uncertainty (see Section 6.2.1), this becomes

$$\Omega_{\text{indistinguishable}} \approx \left[\frac{eV(2mU)^{3/2}}{N^{5/2}h^3} \right]^N \approx \left[\frac{eV(2mU)^{3/2}}{(nN_A)^{5/2} h^3} \right]^{nN_A} \qquad 6.5$$

where again the results are presented both in terms of N molecules and n moles.

This result is very close to the famous result derived by Sackur and Tetrode using more rigorous arguments than the simple ones employed here. The more precise Sackur–Tetrode result is

$$\Omega_{\text{indistinguishable}} \approx \left[\frac{e^{5/2}V\left(4\pi mU/3\right)^{3/2}}{N^{5/2}h^3} \right]^N \qquad 6.6$$

It is now straightforward to obtain the entropy using either Equation 6.5 or 6.6:

$$S = k_B \ln\Omega$$

$$= nk_B N_A \left[\ln\frac{V}{N} + \frac{3}{2}\ln\frac{U}{N} + \text{other constant terms} \right]$$

For one mole $V/N = v$ and $U/N = u$, so

$$s = R\left(\ln v + \frac{3}{2}\ln u + \text{constant terms} \right)$$

From kinetic theory the energy of one mole of monatomic gas is

$$u = \frac{3}{2}N_A k_B T = \frac{3}{2}RT$$

Therefore for a monatomic gas

$$s = R\ln v + \frac{3}{2}R\ln T + s_0 \qquad 5.11$$

where s_0 is a constant. Because $c_v = 3/2R$ for a monatomic gas (Section 3.4.3), this is the identical result to Equation 5.11, which was obtained using macroscopic ideas.

The results of Sections 6.2.4 and 6.2.5 are consistent with the fact that the statistical entropy $S = k_B \ln\Omega$ is the same as the thermodynamic entropy.

The Sackur–Tetrode equation is important historically because it suggests quantization (and a value for Planck's constant h) based solely on considering results from experiments in thermodynamics. The result provides a basis for quantum theory that is independent of Planck's first conception of it, which he developed in the theory of blackbody radiation (Chapter 13), and from Einstein's idea of quantization based on the photoelectric effect.

EXAMPLE 6.2

Use the Sackur–Tetrode equation to compute the entropy of one mole of helium gas at atmospheric pressure and room temperature 293 K. Repeat for one mole of argon gas under the same conditions. *Hint*: Both of these gases are monatomic.

Solution: It will be simplest to use Equation 6.6 with the following parameters for one mole of helium:

$$N = N_A = 6.022 \times 10^{23}$$

$$V = \frac{N_A k_B T}{P} = 0.0240 \text{ m}^3$$

$$U = \frac{3}{2} RT = 3654 \text{ J}$$

$m = 4.003 \text{ u} = 6.647 \times 10^{-27} \text{ kg}$

By Equation 6.6, the entropy of one mole of gas is

$$S = k_B N_A \ln \left[\frac{e^{5/2} V \left(4\pi m U / 3 \right)^{3/2}}{N_A^{5/2} h^3} \right]$$

Inserting all the numerical values gives $S = 125.7$ J/K.

For argon all the parameters are the same except for the atomic mass, which is $39.95 \text{ u} = 6.634 \times 10^{-26}$ kg, about ten times larger than for helium. With the other parameters remaining the same, the molar entropy becomes 154.4 J/K for argon.

The numerical results for helium and argon are both in good agreement with experimental values. It is interesting to see how changing the molecular mass alone caused a change in the entropy. This difference can be traced back to the momentum phase space cells at the root of the theory in this section.

6.2.6 Degradation of Energy and Heat Death

We conclude this section on entropy by discussing the connection between the increase in entropy of the universe associated with an irreversible process and the decrease in the energy that is available for performing work. This can best be seen by considering an example that recalls the macroscopic view of entropy from Chapter 5.

In Figure 6.7(a), a Carnot engine operates between two reservoirs at T_1 and T_0, where the temperature T_0 of the second reservoir is the lowest temperature available. The efficiency of this engine is $\eta = 1 - (T_0 / T_1)$ and so, if heat q is extracted from the hotter reservoir at T_1, the work delivered is $w = q(1 - T_0/T_1)$. For a Carnot engine the process is reversible, and $\Delta S_{universe} = 0$.

Now suppose the same heat transfer q is made into the engine from a second reservoir at a temperature $T_1' < T_1$. As shown in Figure 6.7(b), this is done by allowing the heat q first to be conducted along a metal bar (insulated from its surroundings) from the reservoir at T_1 to the reservoir at T_1' and then being delivered to the engine. The work given out by this combined device is $w' = q(1 - T_0/T_1')$; this is less than w by an amount

$$\Delta w = w - w' = qT_0\left(\frac{1}{T_1'} - \frac{1}{T_1}\right) \qquad 6.7$$

This second process is irreversible, because the conduction of heat along the bar is irreversible. The entropy change of the universe is

$$\Delta S_{universe} = q\left(\frac{1}{T_1'} - \frac{1}{T_1}\right) \qquad 6.8$$

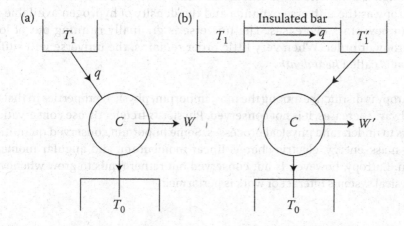

Figure 6.7 An example illustrating the degradation of energy.

because of the entropy changes $-q/T_1$ and $+q/T_1'$ at the two reservoirs. You can see immediately from Equations 6.7 and 6.8 that the amount of work lost by using the second irreversible device is simply

$$\Delta w = T_0 \Delta S_{\text{universe}} \qquad\qquad 6.9$$

Although this point has been illustrated using one example, it can be shown that it is generally true that, in any irreversible process, the energy that becomes unavailable for work is always $T_0 \Delta S_{\text{universe}}$. This result is called the *degradation of energy*. It means simply that the quality (or potential for work) of the energy in the universe decreases by $T_0 \Delta S_{\text{universe}}$ in every irreversible process. This is simply the macroscopic manifestation of the tendency of systems to move into more probable states, as discussed in Section 6.2.2.

It is often said that the world is suffering from an *energy crisis*. You know from the first law that energy is always conserved—energy (or more precisely mass–energy) cannot be destroyed. In the long run, the universe will suffer from an *entropy crisis*. Every irreversible process increases the entropy of the universe, and this means, as was just demonstrated, a loss of capacity of energy for work.

On the scale of Earth, available energy is used by, for example, burning fossil fuels and running nuclear reactors. These fuels are used up because of their entropy increase, as they are converted to less useful molecules and nuclear isotopes. Other sources of energy come directly or indirectly from the sun: photovoltaic cells, wind, and hydroelectric. But eventually the sun will burn out, as its lighter atoms are converted to heavier ones through fusion. The same is true for other stars, and there is also greater entropy as the universe expands and the density of hydrogen available to form new stars decreases. The universe is gradually running out of low entropy or order. When very little order remains, the universe will suffer what is called *heat death*.

Entropy is distinctive among the most important physical properties in that as it always increases, it is not conserved. Physicists frequently use conservation laws to understand physical processes. Some important conserved quantities are mass–energy, electric charge, linear momentum, and angular momentum. Entropy, however, is not conserved but rather tends to grow whenever physical systems interact or work is performed.

6.3 MAXWELL–BOLTZMANN STATISTICS

Much of the foundation for statistical thermodynamics was laid in the middle to late 19th century by James Clerk Maxwell in Britain and Ludwig Boltzmann in Austria. This section introduces the fundamentals of the aptly named Maxwell–Boltzmann statistics.

6.3.1 Boltzmann Factor and Probability

The model shown in Figure 6.8 can be used to derive a general result of great importance in classical statistical physics. This result concerns how the energy of a large, isolated system is partitioned among the various subsystems. In this model, a subsystem of interest labeled A is connected thermally to the rest of the system, a much larger reservoir R. The two subsystems can exchange energy through a diathermal wall, such that the total energy $E_0 = E_A + E_R$ remains constant. The probability of a given energy state E_A is proportional to the total multiplicity of the system and reservoir that gives rise to that macrostate. However, as discussed in Section 6.1, the reservoir's multiplicity of states is much larger than that of the system A, so we can ignore the latter

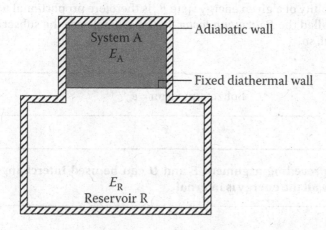

Figure 6.8 A small system A in contact with a large thermal reservoir R. The total energy $E_0 = E_A + E_R$ is constant.

and say that the probability of a state with system energy E_A is proportional to $\Omega(E_R)$, with the stipulation $E_R = E_0 - E_A$. This probability is then proportional to

$$e^{S(E_R)/k_B}$$

where we have used the relation $S = k_B \ln \Omega$.

For the multiplicity $\Omega(E_R)$ the corresponding entropy is $S(E_R) = S(E_0 - E_A)$. Expanding this function in a Taylor series about the total energy E_0,

$$S(E_R) = S(E_0) - E_A \frac{\partial S}{\partial E} + \cdots$$

Neglecting higher-order terms in the expansion and recognizing from the thermodynamic identity that $1/T = \partial S/\partial U$:

$$S(E_R) = S(E_0) - \frac{E_A}{T}$$

Therefore the probability of a state with system energy E_A is proportional to

$$e^{S(E_0) - E_A/k_B T} = e^{S(E_0)} e^{-E_A/k_B T}$$

The probability of a given energy state E_A is therefore proportional to $e^{-E_A/k_B T}$, which is called the *Boltzmann factor*. In the general case the subscript A can be dropped, so

$$\boxed{\text{Boltzmann factor} = e^{-E/k_B T}} \qquad \qquad 6.10$$

In the preceding argument E and U can be used interchangeably, because all the energy is internal.

To find an exact expression for computing probability rather than dealing with simple proportions, remember the approach of flipping coins in Section 6.1. Consider a system with quantized energy levels like the one shown in Figure 6.9. Often quantized energy levels are *degenerate*, meaning that more than one state has the same energy. The quantity called *degeneracy* is defined

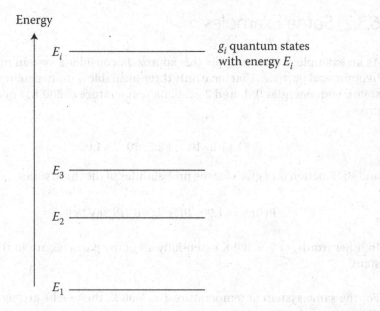

Figure 6.9   Energy levels in a quantum system. Each energy level E_i has a corresponding degeneracy g_i.

as the number (g_i) of states having the same energy (E_i). Taking degeneracy into account, the sum of the Boltzmann factors for all states is defined as the *partition function Z*:

$$Z = \sum_i g_i e^{-E_i/k_B T} \ \text{(partition function)} \qquad 6.11$$

and then the probability of a level with energy E_i is

$$P(E_i) = \frac{1}{Z} g_i e^{-E_i/k_B T} \qquad 6.12$$

You can easily verify that this approach guarantees that the sum of all probabilities is exactly 1. Equation 6.12 is the fundamental result in classical statistical mechanics.

6.3.2 Some Examples

As an example of how to apply this approach, consider a system made up of hypothetical particles that have only three available (non-degenerate) energy states, with energies 0, 1, and 2 eV. For a temperature of 300 K, Equation 6.11 gives

$$Z = 1 + 1.6 \times 10^{-17} + 2.5 \times 10^{-34} \approx 1.00$$

and so Equation 6.12 gives for the probabilities of the three states

$$P(0\,\text{eV}) \approx 1.00 \quad P(1\,\text{eV}) \approx 0 \quad P(2\,\text{eV}) \approx 0$$

In other words, at $T = 300$ K essentially all of the particles are in the ground state.

For the same system at temperature $T = 7500$ K, the results are significantly different:

$$Z = 1 + 0.213 + 0.045 \approx 1.258$$

and

$$P(0\,\text{eV}) \approx 0.79 \quad P(1\,\text{eV}) \approx 0.17 \quad P(2\,\text{eV}) \approx 0.04$$

At higher temperatures there is more thermal energy available to promote some of the particles to higher energy states.

This result of this toy model is characteristic of quantum systems. As an example of a real system, consider hydrogen gas at the surface of a fairly hot star with $T = 7500$ K. For atomic hydrogen, we may take the first three energy levels to be 0, 10.2, and 12.1 eV. It is well known from atomic physics that (excluding spin) hydrogen's ground state is non-degenerate, the first excited level at 10.2 eV contains four states, and the second excited level at 12.1 eV contains nine states. From Equations 6.11 and 6.12, the partition function and probabilities are

$$Z = 1 + 5.6 \times 10^{-7} + 6.7 \times 10^{-8} \approx 1.00$$

and

$$P(0\,\text{eV}) \approx 1.0 \quad P(10.2\,\text{eV}) \approx 5.6 \times 10^{-7} \quad P(12.1\,\text{eV}) \approx 6.7 \times 10^{-8}$$

The 10.2-eV jump from the ground state to the first excited state is large enough compared with $k_B\,T \approx 0.65$ eV that relatively few excited states are populated, even at such high temperatures. However, the probability is large enough that these states are easily detected spectroscopically. Just consider how many atoms are present at the surface of a star!

The mean energy $\bar{E}$ per particle is the average of energy over all the available states, weighted by the appropriate probability factor given by Equation 6.12. That is,

$$\bar{E} = \sum_i E_i P(E_i) \qquad\qquad 6.13$$

For the example presented above with energy states 0, 1, and 2 eV, the mean energy at 7500 K is

$$\bar{E} = \sum_i E_i P(E_i) = 0 + (1\text{eV})(0.17) + (2\text{eV})(0.04) = 0.25 \text{ eV}$$

compared with an average energy close to zero at $T = 300$ K.

6.4 IDEAL GASES

A statistical approach to thermodynamics can be applied effectively to the study of ideal gases. This analysis, originally due to Maxwell, provides an elegant approach to understanding the macroscopic behavior of gases based on microscopic principles.

6.4.1 One-Dimensional Gas

Think of a monatomic gas consisting of identical molecules of mass m that are free to travel only in one dimension, back and forth along the x-axis. Although this model is physically unrealistic, applying statistical methods to a one-dimensional gas yields some interesting results and illustrates the methods that will be needed to study a real gas in three dimensions.

In the classical limit, there are many states with energies E_i that are so closely spaced that it is reasonable to think of a continuous distribution of energies. For the one-dimensional ideal gas, the molecular energy is just the kinetic energy $E = \frac{1}{2}mv_x^2$, where all the molecules in the gas have the same mass m.

For a nearly continuous distribution, the sum in Equation 6.11 can be replaced by an integral:

$$Z = \int_{-\infty}^{\infty} e^{-mv_x^2/2k_BT} dv_x$$

The integral is taken over all possible velocity components v_x, which can be both positive and negative for a gas of molecules free to travel in both directions. This is a standard Gaussian integral with result $Z = \sqrt{2\pi k_B T/m}$. (See Appendix B for discussion of this and other definite integrals of this form.) Therefore by Equation 6.12 the one-dimensional gas follows a velocity distribution of the form

$$f(v_x) = \sqrt{\frac{m}{2\pi k_B T}} e^{-mv_x^2/2k_BT} \qquad 6.14$$

This is a Gaussian distribution with respect to v_x, which perhaps surprisingly has its peak (highest probability) at $v_x = 0$. By symmetry, the mean value of v_x is also zero, which is what might be expected for a gas of molecules that are equally likely to travel in either direction.

More information comes from computing the mean value of the *square* of v_x, which is done by averaging v_x^2 throughout the distribution in Equation 6.14:

$$\overline{v_x^2} = \sqrt{m/2\pi k_B T} \int_{-\infty}^{\infty} v_x^2 e^{-mv_x^2/2k_BT} dv_x$$

This is another standard definite integral, with result

$$\overline{v_x^2} = \frac{kT}{m} \qquad 6.15$$

Two important results follow from Equation 6.15. First, the kinetic energy associated with the mean is $\overline{K} = (1/2)m\overline{v_x^2} = (1/2)k_B T$, in agreement with the equipartition theorem for a one-dimensional gas, which has only one degree of freedom. Second, the mean kinetic energy of a three-dimensional gas follows, because the results for the other directions y and z can be no different:

$$\overline{K} = \frac{1}{2}m\overline{v^2} = \frac{1}{2}m\overline{v_x^2} + \frac{1}{2}m\overline{v_y^2} + \frac{1}{2}m\overline{v_z^2} = 3\left(\frac{1}{2}k_{\mathrm{B}}T\right) = \frac{3}{2}k_{\mathrm{B}}T$$

This matches the prediction of the equipartition theorem (Section 3.4.1) and gives the correct root-mean-square speed of the gas

$$v_{\mathrm{rms}} = \sqrt{\frac{3k_{\mathrm{B}}T}{m}}$$

as given by Equations 3.13 and 3.14.

The velocity distribution for an ideal three-dimensional gas follows from the one-dimensional distribution in Equation 6.14. The three dimensions are indistinguishable from one another. Therefore, the three-dimensional distribution contains three factors, the first given in Equation 6.14 and the other two identical except that they contain v_y and v_z in place of v_x. Thus the three-dimensional distribution is

$$f\left(v_x,v_y,v_z\right) = \left(\frac{m}{2\pi k_{\mathrm{B}}T}\right)^{3/2} e^{-mv_x^2/2k_{\mathrm{B}}T} e^{-mv_y^2/2k_{\mathrm{B}}T} e^{-mv_z^2/2k_{\mathrm{B}}T}$$

which simplifies to

$$f\left(v_x,v_y,v_z\right) = \left(\frac{m}{2\pi k_{\mathrm{B}}T}\right)^{3/2} e^{-mv^2/2k_{\mathrm{B}}T} \qquad 6.16$$

because $v^2 = v_x^2 + v_y^2 + v_z^2$.

6.4.2 Maxwell Speed Distribution

For an ideal gas in three dimensions, the molecules move randomly, and it is useful to find an expression for the distribution of speeds that is analogous to

Equation 6.14 but in terms of the speed v rather than velocity components. The approach is similar to that for the one-dimensional gas in Section 6.4.1.

However, in calculating the partition function it is now necessary to include a degeneracy factor, analogous to the one included for atomic states in Section 6.3.2. This factor arises because for a three-dimensional gas the velocity vector may be considered to lie within a three-dimensional phase space. The "point" of the vector lies on a hypersphere of radius v, and the "surface area" of the hypersphere is $4\pi v^2$, which is the effective degeneracy.

Therefore,

$$Z = 4\pi \int_0^\infty v^2 e^{-mv^2/2k_BT}\, dv$$

The limits on this definite integral are now zero to infinity, because only positive speeds are allowed. Evaluating the integral (see Appendix B),

$$Z = \left(\frac{2\pi k_B T}{m}\right)^{3/2} \tag{6.17}$$

This result allows us to write the Maxwell speed distribution:

Maxwell speed distribution:

$$f(v) = \left(\frac{m}{2\pi k_B T}\right)^{3/2} 4\pi v^2 e^{-mv^2/2k_BT} \tag{6.18}$$

This function is shown in Figure 6.10. Notice that it is no longer symmetrical like the purely Gaussian one-dimensional velocity distribution, but rather rises more quickly at lower speeds and has a longer "tail" at higher speeds. This shape has physical consequences that will be discussed in Section 6.4.3.

6.4.3 Characteristics of Ideal Gases

The Maxwell speed distribution contains a wealth of information that can be mined through statistical analysis. To begin (and to verify that the distribution

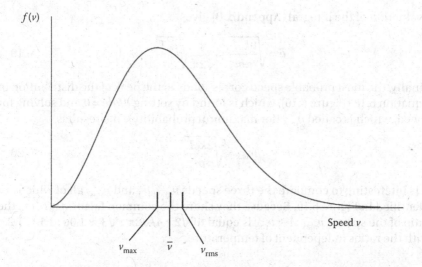

Figure 6.10   Maxwell speed distribution, showing key points on the graph: v_{max}, $\bar{v}$, and v_{rms}.

is correct), the rms speed is found by computing the mean square speed $\overline{v^2}$ directly from the distribution:

$$
\overline{v^2} = \int_0^\infty v^2 \left(\frac{m}{2\pi k_B T}\right)^{3/2} 4\pi v^2 e^{-mv^2/2k_B T} \, dv
$$

$$
= 4\pi \left(\frac{m}{2\pi k_B T}\right)^{3/2} \int_0^\infty v^4 e^{-mv^2/2k_B T} \, dv
$$

This is another standard definite integral (Appendix B), with result

$$
\overline{v^2} = 4\pi \left(\frac{m}{2\pi k_B T}\right)^{3/2} \frac{3}{8}\left(\frac{2k_B T}{m}\right)^2 \sqrt{\frac{2\pi k_B T}{m}} = \frac{3k_B T}{m}
$$

and so $v_{rms} = \sqrt{3k_B T/m}$ as before.

Similarly, the mean speed $\bar{v}$ is

$$
\bar{v} = \int_0^\infty v \left(\frac{m}{2\pi k_B T}\right)^{3/2} 4\pi v^2 e^{-mv^2/2k_B T} \, dv
$$

$$
= 4\pi \left(\frac{m}{2\pi k_B T}\right)^{3/2} \int_0^\infty v^3 e^{-mv^2/2k_B T} \, dv
$$

Evaluation of the integral (Appendix B) gives

$$\bar{v} = \sqrt{\frac{8k_BT}{\pi m}} = \frac{4}{\sqrt{2\pi}}\sqrt{\frac{k_BT}{m}}$$ 6.19

Finally, the most probable speed corresponds to the peak of the distribution in Equation 6.18 (Figure 6.10), which is found by setting $df/dv = 0$ and solving for speed, which is called v_{max} (for maximum probability). The result is

$$v_{max} = \sqrt{\frac{2k_BT}{m}}$$ 6.20

It is interesting to compare the three speeds v_{max}, $\bar{v}$, and v_{rms}, all of which are identified in Figure 6.10. Because they share the common factor $\sqrt{k_BT/m}$, the ratio of the speeds $v_{max} : \bar{v} : v_{rms}$ is equal to $\sqrt{2} : 4/\sqrt{2\pi} : \sqrt{3} \approx 1.00 : 1.13 : 1.22$, with the ratios independent of temperature.

Figure 6.11 shows what happens to the distribution for a given sample of gas as its temperature changes. As temperature increases, the peak shifts to the

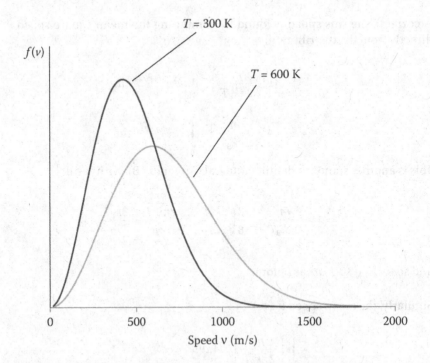

Figure 6.11   Maxwell speed distribution for $T = 300$ K and $T = 600$ K.

right. This is as expected, because temperature is proportional to thermal energy, and the higher thermal energy appears as kinetic energy in the individual molecules. The curve also broadens at higher temperatures, indicating the speeds are distributed more widely. This is in keeping with the constant ratios $v_{max} : \bar{v} : v_{rms}$ discussed above.

Different gases at a given temperature have speeds that vary inversely as $\sqrt{m}$, so in a mixture of gases the lighter ones travel faster on average. For example in air at 273 K, nitrogen molecules (N_2) have v_{rms} = 493 m/s, but for slightly heavier oxygen (O_2) v_{rms} = 461 m/s.

The light noble gas helium, which is monatomic and has a mass of only 4 u, deserves special consideration. At 273 K helium has v_{rms} = 1300 m/s. This is much faster than v_{rms} for the most common gases in air: nitrogen, oxygen, argon, and water vapor. However, 1300 m/s is much less than the escape speed from Earth's surface, which is just over 11 km/s. Why then does helium escape the atmosphere, while the heavier gases remain? Consider the Maxwell speed distribution (Figure 6.10). The fraction of helium atoms traveling faster than the escape speed is large enough that the number dissipates in time. This does not happen, however, for the heavier gases that remain. The loss of helium is a serious issue. Helium is a non-renewable resource (except by renewal in small amounts through alpha decay). It is essential for scientific purposes, such as refrigeration to low temperatures, because of its low boiling point of 4.2 K at atmospheric pressure. Helium also has important medical uses (such as MRI) and industrial applications.

The preceding analysis implies that a planet must have a certain minimum size to retain important atmospheric gases such as oxygen and water vapor. This is a consideration for astrophysicists who study *exoplanets* and explore the possibility of life there. Our moon has no atmosphere, so obviously it falls below that minimum size, even though it is relatively large compared with most natural satellites in our solar system.

> The theory presented here matches experimental results for gases with non-relativistic molecular speeds. When speeds are relativistic, the Maxwell distribution is replaced by the *Maxwell–Jüttner distribution*, which we shall not pursue here.

6.4.4 Distribution of Kinetic Energy

It is sometimes useful to express the Maxwell distribution in terms of energy E rather than velocity or speed. Instead of finding a partition function and degeneracy factor, it is easiest to use the speed distribution in Equation 6.18 and then transform it to an energy distribution $f(E)$ using the fact that

$$f(E)dE = f(v)dv \qquad\qquad 6.21$$

Because $E = \tfrac{1}{2}mv^2$, $dE/dv = mv$, and therefore by Equations 6.18 and 6.21:

$$f(E)=\frac{1}{mv}\left(\frac{m}{2\pi k_B T}\right)^{3/2} 4\pi v^2 e^{-mv^2/2k_B T}$$

which, again using $E = \tfrac{1}{2}mv^2$ simplifies to

$$f(E)=\frac{2}{\sqrt{\pi}}(k_B T)^{-3/2}\, E^{1/2}e^{-E/k_B T} \qquad\qquad 6.22$$

This functional form, with the distribution proportional to $E^{1/2}$ and the Boltzmann factor $e^{-E/k_B T}$, is characteristic of classical distributions.

6.4.5 Maxwell's Demon

In his 1871 book *Theory of Heat*, Maxwell noted that the speed distribution in gases might provide a way to defeat the second law. As an example, he suggested the following strategy, using a vessel filled with air:

> Now let us suppose that such a vessel is divided into two portions, A and B, by a division in which there is a small hole, and that a being, who can see the individual molecules, opens and closes this hole, so as to allow only the swifter molecules to pass from A to B, and only the slower ones to pass from B to A. He will thus, without expenditure of work, raise the temperature of B and lower that of A, in contradiction to the second law of thermodynamics.

Notice that the sorting has lowered the entropy of the universe, which is problematic, because it appears to violate the second law. Any temperature difference created in this manner might, for example, be used to run a heat engine and obtain work without any input of work. Later William Thomson (Lord Kelvin) coined the term *Maxwell's demon* for such a hypothetical device, seizing on the fact that it would be a devilish accomplishment to defeat the second law. Other forms of the demon are easy to imagine. The demon described by Maxwell might pay no attention to speeds but simply allow all the molecules to pass in one direction, filling one half of the vessel. That would also lower the gas's entropy (by doing the reverse of free expansion), and the resulting pressure difference could be used do work at no cost.

Despite these and many other ways people have imagined to use molecular motion to negate the second law, physicists currently believe that the second law is valid. The reasons are subtle and depend on the details of each type of demon. Maxwell's original conception depended on precise measurements of the positions and velocities of individual molecules, so that the hole could be opened and closed in such a way to perform the imagined sorting. The measurement process involves some physical interaction, which may add enough entropy to the system to offset the entropy reduction created by the demon.

Leaving aside the measurement process, there are issues with the information itself. Once the molecules have been measured and sorted, the information about them does not simply go away. Rather, it must be erased, via some other physical process. In the 1960s, Rolf Landauer showed that such erasure of information—necessary to complete a cyclic process—generates enough entropy to offset the demon's gains. The connection between entropy and information is an interesting subject that impacts work on fundamental issues of computing.

In the 21st century, work continues into understanding whether the second law is rigorously followed in all processes, as other kinds of demons are imagined and studied. Some further insight into the rich literature on the subject is found in the compilation by Leff and Rex (2003).

Problems

6.1

    a   Use the thermodynamic identity to show that temperature can be computed using the following expression:

$$\frac{1}{T} = \frac{\partial S}{\partial U}\bigg|_V$$

b   Starting with either Equation 6.5 or 6.6, show that the relation you found in part (a) gives the correct result for the temperature of a monatomic gas. Because of the difficulty involved in defining temperature in some situations, $1/T = \partial S/\partial U$ provides a useful alternative definition, provided the entropy is known as a function of U.

6.2

a   Use the thermodynamic identity to show that pressure can be computed from the entropy function using:

$$\frac{P}{T} = \frac{\partial S}{\partial V}\bigg|_U$$

b   Starting with either Equation 6.5 or 6.6, show that the relation you found in (a) gives the correct result for the pressure of a monatomic gas.

6.3   Verify that the sum of probabilities given by Equation 6.12 is exactly one.

6.4   Using the shorthand notation $\beta = 1/k_B T$, show that the mean energy of a system with partition function Z is

$$\bar{E} = -\frac{1}{Z}\frac{\partial Z}{\partial \beta}$$

6.5   The equipartition theorem says that there is a mean energy $\frac{1}{2} k_B T$ associated with each degree of freedom in a system. To prove this statistically, begin by assuming that the energy for some generalized coordinate x is quadratic and has the form Cx^2 where C is a constant. Then the general form of the partition function is

$$Z = \sum_x e^{-Cx^2/k_B T}$$

a   Assuming the states form a nearly continuous function in x, convert the sum to a definite integral by dividing the domain into small finite intervals Δx. Evaluate the definite integral to find Z as a function of temperature. (b) Using the result of Problem 6.4, show that the mean energy for a single degree of freedom agrees with the equipartition theorem. (c) Show that your result in part (a) is consistent with the partition function $Z = \sqrt{2\pi k_B T/m}$ derived in the text for a one-dimensional gas.

6.6 In quantum mechanics a particle of mass m confined to a one-dimensional infinite potential well of width L has quantized energy states

$$E = \frac{n^2 h^2}{8mL^2}$$

where the quantum number n is restricted to positive integers.

a   Write the partition function for a single particle in one dimension in the form of a sum over all possible quantum numbers n. Change the sum to an integral and thereby show that for this system

$$Z = L\sqrt{\frac{2\pi m k_B T}{h^2}}$$

b   Argue that for a three-dimensional box of volume V, the single-particle partition function becomes

$$Z = V\left(\frac{2\pi m k_B T}{h^2}\right)^{3/2}$$

c   As shown in Section 6.2.5, the result in (b) scales to $Z = \frac{V^N}{N!}$ $\left(\frac{2\pi m k_B T}{h^2}\right)^{3N/2}$ for a gas of N particles. Use the result of Problem 6.4 to show that the mean energy of a particle in this system is $\frac{3}{2}k_B T$, in agreement with the equipartition theorem.

6.7 In purely rotational systems angular momentum L is quantized according in the form

$$L^2 = \ell(\ell+1)\hbar^2$$

where ℓ is a quantum number 0, 1, 2, ... The rotational kinetic energy is $L^2/2I$, where I is rotational inertia. Consider the rotation of a diatomic molecule such as CO. (a) Show that the molecule's rotational energy is quantized in the form $E = \ell(\ell + 1)E_0$, where $E_0 = \hbar^2/2I$. (b) Write the partition function as a sum over angular momentum states ℓ. (Note that the degeneracy of any state ℓ is $2\ell + 1$.) (c) If $k_B T \gg E_0$, it is reasonable to convert the sum to a definite integral. Write the integral and evaluate it. (d) Use the result of Problem 6.4 to find the mean energy of a rotational molecule, valid in the high-temperature limit.

Show that the result is consistent with the equipartition theorem. (e) Explain why the partition function of a homopolar molecule such as O_2 is half the value you found in part (c).

6.8 The molecule CO has a bond length 113 pm. (a) Evaluate the molecule's rotational inertia around an axis through its center of mass and perpendicular to the bond axis. (b) Use the result of Problem 6.7 to find the constant E_0 and partition function Z at $T = 293$ K. Discuss the validity of the approximation $k_B T \gg E_0$ at that temperature.

6.9 Consider a paramagnetic material in which each atom has a magnetic moment in one of two possible orientations, called up and down, which means that each magnetic moment $\vec{\mu}$ points parallel (up) or antiparallel (down) with respect to an applied magnetic field $\vec{B}$. The energy of a magnetic moment in the field is $U = -\vec{\mu} \cdot \vec{B}$. (a) Find the partition function for a single magnetic moment. (b) Find the mean magnetic moment when a magnetic field $\vec{B}$ is applied at temperature T. (c) Explain why your answer to (b) makes sense at extremely low and high temperatures.

6.10 For atomic hydrogen, the allowed energy levels are given by the Bohr equation

$$E_n = -\frac{13.6\text{eV}}{n^2}$$

which gives energies of -13.6, -3.4, and -1.5 eV for the first three energy levels. Rework the example in Section 6.3.2 with atomic hydrogen at 7500 K using these three energy levels. Compute (a) the partition function and (b) the probabilities of the first three levels. (c) Compare your results with the example in the text.

6.11 Show that if quantized energy levels in a system are all changed by an additive constant E_0, then the resulting probabilities given by Equation 6.12 are unchanged.

6.12 Suppose there is a quantized system that can be in one of three energy states, having energies 0, 0.2, and 0.4 eV, respectively. The system is at 5000 K. (a) Compute the partition function for this system. (b) Find the mean energy. (c) Compute the probability that each of the three states will be occupied.

6.13 In his 1884 book *Flatland*, Edwin Abbot dreamed of a two-dimensional world. (a) Find the ideal gas speed distribution analogous to Equation 6.18 for such a world. (b) Find the mean kinetic energy for a monatomic gas as a function of temperature and show that your result is consistent with the equipartition theorem.

6.14 Laser cooling is a technique by which an already cold gas is cooled even further. If a molecule traveling toward a laser absorbs a photon, the collision slows the molecule and effectively cools the gas. In a typical experimental setup, the gas consists of rubidium atoms (m = 85 u), and the laser wavelength is 780 nm. Find the rms speed of a rubidium atom if the effective temperature is (a) 1.0 µK and (b) 1.0 nK. (c) An atom with the rms speed in the 1.0 µK gas absorbs a photon head on. By what fraction is its speed reduced?

6.15 In World War 2, scientists at Oak Ridge, Tennessee (USA) used gaseous diffusion as one method of separating the uranium isotopes 235 and 238, in order to obtain higher concentrations of the fissionable 235 isotope, which has a natural abundance of less than 1%. To accomplish this, natural uranium was put into the form of a gas UF_6 and then allowed to diffuse through a porous barrier. (a) What is the difference in the rms speeds of UF_6 molecules consisting of the two isotopes at $T = 300$ K? (b) Repeat part (a) if the temperature is raised to 800 K.

6.16 (a) For nitrogen molecules at $T = 273$ K, find v_{max}, $\bar{v}$, and v_{rms}. (b) Use numerical integration to find the fraction of nitrogen molecules moving faster than v_{max}, the fraction moving faster than $\bar{v}$, and the fraction moving faster than v_{rms}. (c) Do your results in (b) depend on the type of molecule (monatomic, diatomic, etc.) or the temperature? Explain.

6.17 Temperatures in the thermosphere (in the upper atmosphere) can be 800 K or higher. (a) For a temperature of 800 K, perform a numerical integration to compare the fraction of nitrogen molecules traveling faster than 11 km/s (escape speed). (b) Repeat for helium at the same temperature. (c) Compare the results from (a) and (b) and explain why the atmosphere contains nitrogen but not helium.

6.18 Radioactive isotopes of radon gas are released from materials used in houses. The most common isotope is radon 222, which comes from the decay chain of uranium 238. Radon 222 is an alpha emitter that can cause lung cancer when inhaled in sufficient quantities. Because of its heavy mass, radon gas tends to remain in houses and build in concentration over time. To understand why, compare the rms speed of this (monatomic) radon isotope with nitrogen at $T = 293$ K.

6.19 Use the energy distribution $f(E)$ from Section 6.4.4 to find the mean energy of a molecule in a monatomic ideal gas. Show that your result is consistent with the equipartition theorem.

6.20 Show that each of the following distributions is normalized, that is, that when summed (integrated) over all possible values, they yield a total probability of 1: (a) Maxwell velocity distribution (Equation 6.14); (b) three-dimensional velocity distribution (Equation 6.16); (c) Maxwell speed distribution (Equation 6.18); and (d) energy distribution (Equation 6.22).

6.21 (a) Use the Sackur–Tetrode equation to find the molar entropy of neon. Compare with the experimental value of 146 J/K. (b) Repeat for radon, which has a measured molar entropy 176 J/K. *Note*: both neon and radon are monatomic gases.

REFERENCES

Leff, H.S. and Rex, A.F., *Maxwell's Demon 2*, IOP Publishing, Bristol, 2003.
Maxwell, J.C., *Theory of Heat*, Longmans, Green, & Co., London, 1871.

Chapter 7: The Thermodynamic Potentials and the Maxwell Relations

This chapter introduces a number of new quantities, in particular the so-called thermodynamic potentials, which are needed to provide important links between theoretical and experimental work. This is all based on previous chapters and concepts with which you are already familiar, so a brief review is in order.

7.1 THERMODYNAMIC POTENTIALS

In the formulation of thermodynamics to this point, you have seen how the first law makes it possible to define the internal energy U as the sum of the random kinetic and potential energies of the component particles of the system. The second law allowed a definition of the entropy S. The two laws were combined into the thermodynamic identity

$$TdS = dU + PdV \qquad 7.1$$

where it is assumed that the only work done is due to a change in volume. This equation is identically true in that it holds for both reversible and irreversible infinitesimal processes. The power of this equation has been evident in the preceding chapters and will be again in this and the following chapters. However, although its physical interpretation is clear, U is not well suited for the analysis of certain thermodynamic processes. It is convenient to introduce three additional state functions, closely related to U, all of which have the dimensions of energy. These functions are the enthalpy H, which was defined in Chapter 3; the *Helmholtz function F*; and the *Gibbs function G*. These functions provide a more direct link with experiments than can be obtained with the use of U alone. There is a fifth function, the chemical potential, μ, which is useful in discussing the thermodynamics of open systems where the mass

DOI: 10.1201/9781003299479-7

of the system is not constant; however, the discussion of μ will be deferred until Chapter 11. The four functions U, H, F, and G have a wide applicability throughout thermodynamics, and as a set they are referred to as the four *thermodynamic potentials*. The next few sections present their properties in turn. In the course of this discussion, we present four extremely useful general thermodynamic relations among the four variables P, V, T, and S—the four Maxwell relations.

7.2 INTERNAL ENERGY U

Equation 7.1 gives for the internal energy U

$$dU = TdS - PdV \qquad\qquad 7.2$$

As long as work is being done only through a $-PdV$ process, this equation is independent of the type of process used, and so any relations obtained from it are general ones.

The form of Equation 7.2 suggests that U may be written in terms of the independent pair of variables S and V as $U = U(S, V)$, where the notation $U(S, V)$ denotes a function of S and V. Hence,

$$dU = \left(\frac{\partial U}{\partial S}\right)_V dS + \left(\frac{\partial U}{\partial V}\right)_S dV \qquad\qquad 7.3$$

Comparing Equations 7.2 and 7.3,

$$T = \left(\frac{\partial U}{\partial S}\right)_V \quad \text{and} \quad P = -\left(\frac{\partial U}{\partial V}\right)_S \qquad\qquad 7.4$$

This means that, if U can be expressed in terms of its so-called *natural variables* V and S, then these derivatives allow you to find T and P. Further, because U is a state function, dU is an exact differential. Using the condition for a differential to be exact (Equation B.15 in Appendix B) in Equation 7.2

$$\left(\frac{\partial T}{\partial V}\right)_S = -\left(\frac{\partial P}{\partial S}\right)_V \qquad\qquad 7.5$$

This is the first so-called *Maxwell relation*. Notice that the natural variables of U, namely S and V, are the quantities appearing outside the partial derivatives.

The results of this section can be used to derive two useful expressions for the heat capacity at constant volume $C_V = đQ_V/dT$, where the heat has to be put in reversibly. It follows from Equation 7.2 that in a constant volume (or isochoric) process

$$TdS = dU \quad \text{(isochoric)} \qquad\qquad 7.6$$

Also, $TdS = đQ$ for a reversible process, and it follows from this and Equation 7.6 that for a reversible isochoric process

$$đQ_V = dU \quad \text{(reversible and isochoric)} \qquad\qquad 7.7$$

Hence,

$$C_V = \left(\frac{\partial U}{\partial T}\right)_V \qquad\qquad 7.8$$

and

$$C_V = T\left(\frac{\partial S}{\partial T}\right)_V \qquad\qquad 7.9$$

Notice that Equation 7.8 agrees with Equation 3.6, which was derived using the first law.

7.3 ENTHALPY *H*

In Chapter 3, enthalpy H was defined as

$$H = U + PV \qquad\qquad 3.7$$

This is a state function, because all the quantities on the right side take unique values for each state. Differentiating,

$$dH = dU + PdV + VdP \qquad\qquad 7.10$$

Using Equation 7.1

$$dH = TdS + VdP \qquad\qquad 7.11$$

This equation holds for both reversible and irreversible processes by the same argument that was used to show that the thermodynamic identity was independent of the type of process, namely that the equation involves only state functions on each side. Again, as this equation is independent of the type of process, any relations obtained from it are general ones.

Following the same process as in Section 7.2 for internal energy U, it is useful to write $H = H(S, P)$ and evaluate the exact differential:

$$dH = \left(\frac{\partial H}{\partial S}\right)_P dS + \left(\frac{\partial H}{\partial P}\right)_S dP \qquad\qquad 7.12$$

Comparing this equation with Equation 7.11,

$$T = \left(\frac{\partial H}{\partial S}\right)_P \quad \text{and} \quad V = \left(\frac{\partial H}{\partial P}\right)_S \qquad\qquad 7.13$$

This means that, if you know H in terms of its natural variables S and P, you can find both the temperature and the volume. Further, using the condition for dH in Equation 7.11 to be an exact differential,

$$\left(\frac{\partial T}{\partial P}\right)_S = \left(\frac{\partial V}{\partial S}\right)_P \qquad\qquad 7.14$$

This is the second Maxwell relation, with the natural variables S and P appearing outside the partial derivatives.

The most important property of enthalpy is that the change in H is equal to the heat flow in an isobaric, reversible process. This result was found in Chapter 3 in the infinitesimal form as Equation 3.9

$$đQ_P = dH \quad \text{(isobaric and reversible)} \qquad\qquad 7.15$$

This result can also be obtained directly from Equation 7.11, which gives $dH = TdS$ for an isobaric process. Because $TdS = đQ$ for a reversible process, Equation 7.15 then follows.

There are two useful expressions for the heat capacity at constant pressure, $C_P = đQ_P/dT$ where the heat has to be added reversibly. From Equation 7.15,

$$C_P = \left(\frac{\partial H}{\partial T}\right)_P \qquad\qquad 7.16$$

Also, because $T\,dS = đQ$ for a reversible process,

$$C_P = T\left(\frac{\partial S}{\partial T}\right)_P \qquad\qquad 7.17$$

Notice that Equation 7.16 agrees with Equation 3.10, which was derived using the first law.

In Section 3.3.1, it was noted that the enthalpy change in a chemical reaction taking place under constant external pressure is equal to the heat of reaction. This is the situation that normally holds in chemistry because most reactions occur under constant atmospheric pressure. The important point is that this result holds for a chemical reaction, even though this change is irreversible in the thermodynamic sense in that the system does not go through a series of equilibrium states. Of course, a chemical reaction can be made to go the other way, but this usually requires a finite (as opposed to an infinitesimal) change in the external conditions of temperature and possibly pressure. Therefore, chemical reactions cannot in general be regarded as thermodynamically reversible.

To see this, imagine the chemicals being contained in a cylinder fitted with the usual light, frictionless piston as in Figure 7.1. Suppose the volume of the chemicals changes by ΔV, say by a gas being produced, and that the heat of reaction Q is given to the chemical system. This will push the piston back

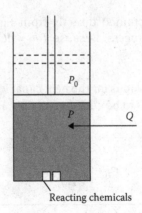

P_0

P

Q

Reacting chemicals

Figure 7.1   A chemical reaction occurring under conditions of constant external pressure. The heat of reaction is $Q = \Delta H$.

against atmospheric pressure P_0 until it comes to rest again when the reaction is over. The external work done by the system is then $P_0 \Delta V$. Applying the first law to this system,

$$\Delta U = Q - P_0 \Delta V \qquad 7.18$$

so

$$Q = \Delta U + P_0 \Delta V = \Delta U + \Delta(PV) \qquad 7.19$$

The last step in Equation 7.19 follows from the fact that the pressure $P = P_0$ at the start and end of the process, even though it may vary during the reaction. It follows immediately that

$$Q = \Delta H \qquad \text{(constant pressure)} \qquad 7.20$$

which is the desired result. Notice that the only restriction made was the one of constant external pressure.

This important result is often developed erroneously in the following manner. Equation 7.11 gives for an infinitesimal process

$$dH = TdS + VdP$$

If this process is reversible and isobaric so that $TdS = dQ$ and $dP = 0$, then $dH = dQ$ and $\Delta H = Q$ for a finite process, which appears to be the required result. However, as explained above, chemical reactions are not generally reversible, and so this argument is inapplicable.

7.4 HELMHOLTZ FUNCTION *F*

This state function is designed for problems in which temperature and volume are the important variables and is also of value in statistical mechanics. It is defined as

---

Helmholtz function *F*:

$$F = U - TS \qquad\qquad 7.21$$

For an infinitesimal change,

$$dF = dU - TdS - SdT \qquad\qquad 7.22$$

Using Equation 7.1,

$$dF = -PdV - SdT \qquad\qquad 7.23$$

Again, this equation is independent of the type of process, so any relations obtained from it are general ones. The form of Equation 7.23 suggests that the natural variables of F are V and T, so symbolically $F = F(V, T)$. Hence,

$$dF = \left(\frac{\partial F}{\partial V}\right)_T dV + \left(\frac{\partial F}{\partial T}\right)_V dT \qquad\qquad 7.24$$

Comparing coefficients in Equations 7.23 and 7.24,

$$P = -\left(\frac{\partial F}{\partial V}\right)_T \quad\text{and}\quad S = -\left(\frac{\partial F}{\partial T}\right)_V \qquad\qquad 7.25$$

Hence, if you know F as a function of volume and temperature, you can find both the entropy and the pressure.

Because F is a function of state, dF is an exact differential. The condition for an exact differential in Equation 7.23 gives

$$\left(\frac{\partial P}{\partial T}\right)_V = \left(\frac{\partial S}{\partial V}\right)_T \qquad\qquad 7.26$$

This is the third Maxwell relation, with the natural variables V and T appearing outside the derivatives.

There are some additional properties of F, discussed in the sections that follow in the remainder of Section 7.4.

7.4.1 Maximum Work from a System with $\Delta T = 0$

In a purely mechanical system with no thermal considerations (e.g., two point masses attracted by gravity), the principle of conservation of mechanical energy says that the work performed by the system is equal to the decrease in potential energy. In thermodynamics, however, the situation is complicated by the fact that energy can also be exchanged between the system and the surroundings in the form of heat, and so there is more to the relation between the work performed and the change of energy. Fortunately, there is a simple expression for the work performed by a system that is in thermal contact with the surroundings at T_0. This means that the end points of the process are at T_0, although the intermediate states traversed by the system are not necessarily at T_0.

Suppose the system is in thermal contact with a heat reservoir at T_0, as in Figure 7.2, and let heat Q pass from the reservoir into the system. The system and reservoir are surrounded by an adiabatic wall to exclude any other heat flow. The system may perform work W; this may be volume work because its walls are not necessarily rigid, or work in another form such as electrical work. The principle of increasing entropy gives

$$\Delta S + \Delta S_0 \geq 0 \qquad\qquad 7.27$$

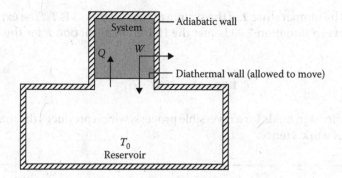

Figure 7.2   A system in thermal contact with a reservoir at T_0. The system performs work W equal to the decrease in F.

where ΔS and ΔS_0 are the entropy changes of the system and the reservoir. But because the temperature of the reservoir is essentially unchanged at T_0, $\Delta S_0 = -Q/T_0$, and Equation 7.27 becomes

$$\frac{\Delta S - Q}{T_0} \geq 0$$

or

$$Q - T_0 \Delta S \leq 0 \qquad\qquad 7.28$$

This is a familiar result, having appeared in differential form as Equation 5.7. Applying the first law to the system,

$$\Delta U = Q - W \qquad\qquad 7.29$$

Equation 7.29 contains a minus sign on the work term, because, as noted above, this analysis concerns work done *by* the system, which is the opposite of the work done *on* the system.

Substituting for Q, given by Equation 7.29, into Equation 7.28

$$\Delta U + W - T_0 \Delta S \leq 0$$

or

$$\Delta(U - TS) + W \leq 0 \qquad\qquad 7.30$$

because the temperature T of the system at the end points is T_0. The expression in brackets in Equation 7.30 is just the Helmholtz function F for the system, and so

$$W \leq -\Delta F \quad (\Delta T = 0) \qquad 7.31$$

The equality sign holds for a reversible process, which produces the maximum amount of work. Hence,

> **In a process in which the end point temperatures of a system are the same as those of its constant-temperature surroundings, the maximum work obtainable is equal to the decrease in the Helmholtz function of the system.**

Such processes are commonly encountered. Alternately, F is the amount of work that must be done to create a system out of nothing at constant temperature. This can be seen by Equation 7.21, where U is the internal energy of the system created and TS is the heat that can be absorbed "for free" from the environment at temperature T. Because of this interpretation, F is often called the *Helmholtz free energy*.

> **In some other books, you will see the Helmholtz function given the symbol A after the German word for work, *Arbeit*. In Section 7.6, we will use the symbol A for a different function, called availability.**

It should be emphasized that T does not have to be held constant in an isothermal process at T_0 for Equation 7.31 to hold but rather has to assume this value only at the end points. A less general proof of Equation 7.31 follows immediately from Equation 7.23 for the special case of an isothermal process. In that case $-dF = PdV = dW_{rev}$.

7.4.2 Equilibrium Condition for a System Held at Constant Volume and Temperature

There is a simple argument to show that the equilibrium condition for a system in thermal contact with a heat reservoir and held at constant volume is

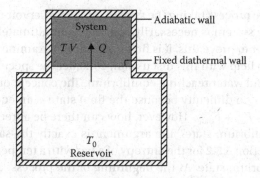

Figure 7.3 A system in thermal contact with a reservoir at T_0. The volume of the system is fixed. The condition for equilibrium is that F is a minimum.

one of minimum F. To see this, consider the system shown in Figure 7.3. The system is in thermal contact, via a diathermal wall, with a reservoir at a temperature T_0, so that its temperature T is also T_0. Let the volume of the system be fixed at V. The combined system of the system and the reservoir is surrounded by an adiabatic wall, as in the previous section.

If the system is a simple one consisting of a fixed mass of homogeneous material, then specifying the temperature and volume of the system fixes its state, and there is complete or thermodynamic equilibrium; clearly, no further changes can occur. However, other internal degrees of freedom may exist within a more complex system, so that it may not be in a state of thermodynamic equilibrium even though the temperature and the volume have been specified. An irreversible process can then occur involving the transfer of heat Q from the reservoir to the system as it tends toward equilibrium.

As an example of what is meant by an extra degree of freedom, imagine a chemical reaction occurring within the system

$$A + B \rightarrow AB$$

with the ratio of the amount of product AB to the amount of reactant A or B being a variable quantity that changes as equilibrium is approached. Alternatively, there could be a mixture of ice and water within the system, with the ratio of ice to water varying until equilibrium is reached. (Strictly speaking, the volume of this system may change, but normally only by a small amount.) In both of these examples, heat is transferred between the system and the reservoir; this is the heat of reaction for the chemical reaction and the latent heat for the melting of the ice.

In an irreversible process taking the system and the reservoir toward equilibrium, ΔF for the system is necessarily negative (and ultimately zero at equilibrium). In order to prove this, it is first necessary to examine exactly what is meant by ΔF. To help with this discussion, consider the specific example of a mixture of ice and water reaching equilibrium. The concept of a well-defined final F causes us no difficulty because the final state is an equilibrium state and the total $F = F_{ice} + F_{water}$. However, how can there be a free energy for the initial non-equilibrium state? The argument is exactly the same as the one employed in Section 5.3.2 for the entropy of a bar with a temperature gradient in a non-equilibrium state. At the beginning of the process, and also at any stage, imagine separating the ice and water so that there is again an equilibrium situation. One may then add $F = F_{ice} + F_{water}$ as before to give a definite F for the system. In this way, ΔF for the process is defined.

Why then is $\Delta F \le 0$? To see this, first assume that the process takes place isothermally at T_0 and at the constant volume V. As the combined system of the system and the thermal reservoir is thermally isolated from the rest of the universe by the adiabatic wall, it follows from the principle of increasing entropy and from Equation 7.28 that $Q - T_0\Delta S \le 0$, where ΔS is the entropy change of the system. Applying the first law to the system, and remembering that no work is done because the volume is constant at V,

$$\Delta U = Q \qquad\qquad 7.32$$

> **This discussion excludes any other types of work such as magnetic work.**

Substituting for Q in Equation 7.28,

$$\Delta U - T_0\Delta S \le 0$$

or

$$\Delta U - \Delta(TS) \le 0 \qquad\qquad 7.33$$

because the temperature T is fixed at the constant value of T_0 for the process under consideration. Rewriting Equation 7.33,

$$\Delta(U - TS) \le 0$$

or

$$\Delta F \leq 0 \qquad\qquad 7.34$$

As always, because F is a state function, the decrease ΔF in the Helmholtz function is the same for any process between a given pair of equilibrium states. Thus, Equation 7.34 holds in general for any process between a pair of states at the same temperature and volume, and not just for the special isochoric and isothermal process just considered. It is important to realize that the temperature and volume do not have to be fixed at T_0 and V during the process for Equation 7.34 to hold, but that they have these values only at the end points. Indeed, in the example of a chemical reaction, the temperature will almost certainly change during the reaction before settling back to T_0 when it is over.

Equation 7.34 shows that, for a system in thermal contact with a heat reservoir and in which the volume is held constant, spontaneous changes in the system (changes that occur of their own accord without the further influence of an agency external to the system and the reservoir) occur in the direction of decreasing F. Eventually, the system will reach thermodynamic equilibrium. There are then no further finite changes in the state variables of the system, and F will reach a minimum. As equilibrium is approached, the process becomes reversible (no finite temperature differences between the system and the surroundings), and $dF = 0$. This is allowed by the equality part of Equation 7.34. Hence,

---

The condition for thermodynamic equilibrium in a system in thermal contact with a heat reservoir and maintained at constant volume is that the Helmholtz function is a minimum.

Finally, there are two important points to remember. The first is that the actual non-thermal restriction imposed on the system is that it is not allowed to do any work on the surroundings. Holding V constant for a PVT system ensures that this is so. The second is that minimizing F for the system is done by maximizing the entropy of the system and the reservoir. A change in F of the system automatically looks after the changes in S for both the system and the reservoir when the two are in thermal contact and the volume is fixed.

As an illustration of these ideas, consider the formation of vacancies in a solid. A simple model of a crystalline solid is a regular array of atoms such as shown in Figure 7.4, which shows the atoms to be arranged in a cubic array.

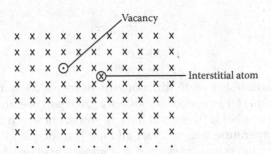

Figure 7.4 Vacancy formation in a crystal.

(Polonium is a pure element that takes this form.) Suppose·that the crystal is in thermal contact with the surroundings at a temperature T, and the volume V of the crystal is fixed. Now, it is possible for one of the atoms, if it gains enough thermal energy, to break free from the bonds of its neighbors and to migrate to a new position in the crystal, leaving behind a vacancy. The migrating atom will sit in a new position either in an interstitial position between two other atoms or will fill a vacancy site that has previously been vacated by another atom. The question is "How many vacancies are to be expected at thermal equilibrium at a given temperature?" Because it costs energy to break the bonds and for each vacancy to be formed, an initial answer to the question is that there should be no vacancies at all, as this will minimize the internal energy U. However, as has just been shown, it is not internal energy U that is minimized, but rather the Helmholtz function $F = U - TS$ of the crystal, consistent with maximizing the entropy of the system and its surroundings. Every time a vacancy is formed, S increases and it is possible to calculate the entropy of a set of n vacancies sitting in the crystal. (See Guenault (2007) Section 10.3 for discussion.) Thus, any increase in U resulting from the formation of vacancies will eventually be more than beaten by the TS term. At a high enough temperature, the result is a lowering of F. This means that vacancies will occur.

7.4.3 Bridge between Thermodynamics and Statistical Mechanics

It is through the Helmholtz free energy that an important link between statistical mechanics and thermodynamics is made. Recall from Chapter 6 (Equation 6.11) that the partition function is

$$Z = \sum_i g_i e^{-E_i/k_B T} \qquad 7.35$$

where g_i is the degeneracy of a quantum level with energy E_i. The Helmholtz function can be written in terms of the single-particle partition function as

$$F = -Nk_BT \ln Z \qquad\qquad 7.36$$

for the N particles making up the system, providing that they are only weakly interacting and are distinguishable from each other. (We will not prove this result, but it is proven elsewhere. See for example Guenault (2007) Section 2.5. Another version of the proof is left to Problem 7.17 at the end of this chapter.) This important relation is known as the *bridge equation* for the so-called canonical distribution in statistical mechanics. Once the partition function has been evaluated, F can be obtained using Equation 7.36. From this, the state functions S and P can be evaluated using Equation 7.25. The thermodynamic description of the system is then known.

As a simple illustration of these ideas, consider a two-level system of N weakly interacting distinguishable particles, where each particle can exist in one of two non-degenerate quantum states with energies 0 and ε. (For example, such a system could be a collection of electron spins, as in a paramagnetic salt, in a magnetic field.) Then,

$$Z = e^{-0/k_BT} + e^{-\varepsilon/k_BT} = 1 + e^{-\varepsilon/k_BT}$$

So by Equation 7.36,

$$F = -Nk_BT \ln\left(1 + e^{-\varepsilon/k_BT}\right)$$

Thus,

$$S = -\left(\frac{\partial F}{\partial T}\right)_V = Nk_B\left[\ln\left(1 + e^{-\varepsilon/k_BT}\right) + \frac{\varepsilon}{k_BT}\cdot\frac{1}{\left(1 + e^{\varepsilon/k_BT}\right)}\right]$$

which gives the variation of entropy with temperature. At low temperatures ($k_B T \ll \varepsilon$), this expression approximates to $S = 0$, while at high temperatures ($k_BT \gg \varepsilon$), it yields $S = Nk_B \ln 2$. These results are consistent with Equation 6.3. In the high-temperature limit, each particle has sufficient thermal energy to be able to occupy either of the two energy levels, and so $\Omega = 2^N$, giving the $S = Nk_B \ln 2$ result again. Also, in the low-temperature limit, each particle can occupy only the lowest level, so $\Omega = 1^N = 1$ with $S = 0$. The statistical (partition function) approach has done much more than this; it has given the actual temperature variation of S.

In general, the partition function is an invaluable tool in the calculation of the bulk thermodynamic properties of a system using statistical mechanics, especially when more complicated systems are considered than the one here.

7.5 GIBBS FUNCTION G

This state function is designed for use in problems where pressure and temperature are the important variables. It is of enormous importance in chemistry and in the study of systems where there is a mixture of two phases of a substance, such as a mixture of ice and water; these systems will be discussed in Chapter 10. The Gibbs function is defined as

---

Gibbs function G:

$$G = H - TS \qquad\qquad 7.37$$

For an infinitesimal change,

$$dG = dH - TdS - SdT$$

Using Equation 3.7, this becomes

$$dG = dU + PdV + VdP - TdS - SdT$$

It follows from Equation 7.1 that

$$dG = VdP - SdT \qquad\qquad 7.38$$

Again this equation is independent of whether or not the process is reversible and so any relations derived from it are general ones. The natural variables of G are P and T. Equation 7.38 shows that, in any process that takes place at constant T and P, the Gibbs function is unchanged.

The form of Equation 7.38 suggests that $G = G(P, T)$. Hence,

$$dG = \left(\frac{\partial G}{\partial P}\right)_T dP + \left(\frac{\partial G}{\partial T}\right)_P dT \qquad\qquad 7.39$$

Comparing the coefficients in Equations 7.38 and 7.39,

$$V = \left(\frac{\partial G}{\partial P}\right)_T \quad \text{and} \quad S = -\left(\frac{\partial G}{\partial T}\right)_P \qquad 7.40$$

Hence, if you know G as a function of P and T, you can find the volume and the entropy.

As G is a state function, dG is an exact differential. Using the condition for an exact differential in Equation 7.38,

$$\left(\frac{\partial V}{\partial T}\right)_P = -\left(\frac{\partial S}{\partial P}\right)_T \qquad 7.41$$

This is the fourth Maxwell relation, with the natural variables P and T appearing outside the partial derivatives. There are some further important properties of G, which are detailed in the following sections.

7.5.1 Equilibrium Condition for a System Held at Constant Pressure and Temperature

In Section 7.4, you saw that a system held at constant volume and in thermal contact with a heat reservoir assumes an equilibrium state which is one of minimum F. Now, it will be shown, again using the principle of increasing entropy, that a system in thermal and mechanical contact with a heat and pressure reservoir moves to an equilibrium state of minimum G.

The reservoir employed for this purpose must be so large that its temperature and pressure remain unchanged, whatever is done to it. This result for the minimization of G is of enormous importance, because the conditions described are exactly the ones encountered in many natural processes, where the surrounding atmosphere at P_0 and T_0 acts as a pressure and heat reservoir. Examples of such processes include most chemical reactions and phase changes.

Consider the system shown in Figure 7.5, which is in contact with a heat and pressure reservoir at T_0 and P_0 via a diathermal piston that is weightless, free,

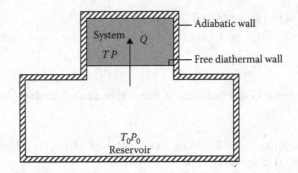

Figure 7.5 A system in contact with a reservoir at T_0 and P_0. The pressure and temperature of the system at the end points are maintained at P_0 and T_0 because the intervening wall is diathermal and can move freely. The condition for equilibrium is that G is a minimum.

and frictionless. As in the previous discussion of F, let there be some internal degree of freedom in the system so that it is not in thermodynamic equilibrium. As before, let heat Q pass from the reservoir to the system in a spontaneous process from an initial non-equilibrium state at P_0 and T_0 to a final equilibrium state also at P_0 and T_0. The change ΔG in the Gibbs function is defined in exactly the same way as ΔF in Section 7.4.2. The following argument will show that ΔG must be negative if the process is irreversible.

First consider what would happen if this process were both isothermal and isobaric. Because the combined system of the system and the reservoir is thermally isolated from the rest of the universe, it follows from the principle of increasing entropy, exactly as in Section 7.4, that $Q - T_0 \Delta S \leq 0$ (Equation 7.28), where ΔS is the entropy change of the system. Applying the first law to the system,

$$\Delta U = Q - P_0 \Delta V \qquad\qquad 7.42$$

where $W = -P_0 \Delta V$, excluding other types of work, such as magnetic work. Substituting for Q using Equation 7.28,

$$\Delta U + P_0 \Delta V - T_0 \Delta S \leq 0$$

or

$$\Delta U + \Delta(PV) - \Delta(TS) \leq 0 \qquad\qquad 7.43$$

because the pressure and temperature of the system are fixed at the constant values of P_0 and T_0 for the process under consideration. Rewriting Equation 7.43,

$$\Delta(U + PV - TS) \leq 0 \qquad 7.44$$

or

$$\Delta G \leq 0 \qquad 7.45$$

The argument as presented holds for an isothermal, isobaric process. However, because G is a state function, the decrease in G is the same for any process between this pair of end point states as it is for the isothermal, isobaric process. Thus, Equation 7.45 holds in general for a process between a pair of states at the same temperature and pressure as the surroundings. It is important to realize that the temperature and pressure do not have to remain fixed at T_0 and P_0 during the process for $\Delta G \leq 0$ to hold, only that they assume these values at the end points. Indeed, in a chemical reaction, the temperature, and possibly the pressure, will almost certainly change before setting back to the ambient temperature and pressure at the final equilibrium state.

Equation 7.45 says that, for a system in contact with a heat and pressure reservoir, spontaneous changes will occur in the direction of decreasing G. At equilibrium, where any infinitesimal changes induced are reversible, the equality sign holds and

$$dG = 0 \qquad 7.46$$

so there is a minimum in the Gibbs function. Hence,

> **The condition for thermodynamic equilibrium in a system in thermal and mechanical contact with a heat and pressure reservoir is that the Gibbs function is a minimum.**

7.5.2 Application to Chemical Reactions

This result (minimizing the Gibbs function at equilibrium when the temperature and pressure are fixed) is significant for chemical reactions. In a typical reaction

$$A + B \rightarrow C + D$$

the reactants A and B are initially at room or ambient temperature. They react, perhaps giving off or absorbing heat, with the temperature rising or falling, and then the products of the reaction C and D return to room temperature. The pressure is kept at the pressure of the surroundings, usually atmospheric pressure as noted before. The reaction will proceed if ΔG for the process is negative. Consider two important biochemical reactions.

The oxidation of glucose is represented by the following reaction:

$$C_6H_{12}O_6 + 6O_2 \rightarrow 6CO_2 + 6H_2O$$

The values for the Gibbs function for all the chemicals are tabulated. At 25°C, the change in G for the reaction is −2870 kJ/mol, so the reaction will proceed in the direction indicated. However, thermodynamics does not give the rate at which the reaction proceeds, only its direction. In practice, an enzyme catalyst is required to accelerate this reaction.

Photosynthesis, the formation of carbohydrate, is represented by the following reaction:

$$CO_2 + H_2O \rightarrow CH_2O + O_2$$

ΔG for this reaction is +478 kJ/mol. With $\Delta G > 0$, this reaction will not proceed spontaneously. The radiant energy of sunlight is required to drive the reaction forward.

EXAMPLE 7.1

Use the tabulated data below to compute ΔG for photosynthesis. The quantities in the table are all for one mole of each substance at $T = 298$ K, so assume a constant temperature $T = 298$ K for the reaction. Compare your result with the ΔG quoted in the text.

Substance	Enthalpy of Formation (kJ)	Entropy (J/K)
CO_2	−393.5	213.74
H_2O	−241.82	69.91
CH_2O	−115.9	218.95
O_2	0	205.14

Solution: With all reaction parameters given at $T = 298$ K, assume that is the constant temperature for the reaction. Then from Equation 7.37, ΔG is given by $\Delta G = \Delta H - T\Delta S$. Using the tabulated values for one mole,

$$\Delta H = -115.9 \text{ kJ} - (-393.5 \text{ kJ} - 241.82 \text{ kJ}) = 519.42 \text{ kJ}$$

$$\Delta S = 218.95 \text{ J/K} + 205.14 \text{ J/K} - (213.74 \text{ J/K} + 69.91 \text{ kJ})$$

$$= 140.44 \text{ J/K} = 0.14044 \text{ kJ/K}$$

$$\Delta G = \Delta H - T\Delta S = 519.42 \text{ kJ} - (298 \text{ K})(-0.14044 \text{ kJ/K}) = 477.6 \text{ kJ}$$

This agrees with the value for ΔG quoted in the text.

7.5.3 Gibbs Function and Entropy

It is instructive to take a closer look, in rather more general terms, at the underlying reasons for the minimization of G for a closed system in thermal and mechanical contact with the surroundings at T_0 and P_0. The second law requires that the entropy of the system and the surroundings can only increase or stay the same

$$\Delta S_{\text{universe}} = \Delta S + \Delta S_0 \geq 0$$

As in Equation 7.43, this implies

$$\boxed{\Delta H - T_0 \Delta S \leq 0}$$

because $\Delta U + \Delta(PV) = \Delta H$ and $\Delta(TS) = T_0\Delta S$. Although the state functions S and H in this inequality refer to the system, it is important to remember that the inequality tells us about the net entropy change in the system and the surroundings. Now, the interdependent entropy and enthalpy changes ΔH and ΔS of the system that occur in a process may be positive or negative, and the necessary requirement for the process to proceed is that the inequality is satisfied so that the entropy of the universe increases. Clearly, this is so if ΔS is positive and ΔH is negative; such a process may then proceed of its own accord (i.e., spontaneously). Conversely, if ΔH is positive and ΔS is negative, the process cannot proceed. However, if ΔH and ΔS have the same sign, then whether or not the process proceeds depends on which of the two terms, ΔH or $T_0\Delta S$ is dominant. There is competition between them, and a balance will be struck when $T_0\Delta S = \Delta H$ with $\Delta S_{\text{universe}} = 0$. This balance at equilibrium appears as a minimization of G, which contains both H and S.

These ideas are in contrast with those from statics, where at equilibrium it is the simple potential energy that is minimized. In thermodynamics, one must allow for the additional complication of adding energy to the system in the form of heat rather than just by work. The potential function G accounts for the competing changes in H and S and automatically includes the entropy changes of both the system and the surroundings for a process in which the end points are at the ambient T_0 and P_0. Section 7.4 considered the less general case of a system at a fixed volume in thermal contact with a heat reservoir at T_0. In that case, there is competition between ΔU and $T_0 \Delta S$ to give a minimum in F at equilibrium.

It is worth examining in more detail the case when the two terms ΔH and $T_0 \Delta S$ have the same sign. Whether or not the process proceeds depends on which term is dominant in the inequality $\Delta G = \Delta H - T_0 \Delta S \leq 0$. Notice that, at high temperatures, the $T_0 \Delta S$ term is dominant, and at low temperatures, the enthalpy term is dominant. Let the process be represented by

$$A \leftrightarrow B$$

where the double arrow implies that the process between two states A and B can go either direction under suitable conditions. Suppose that the process $A \to B$ has $\Delta S > 0$, which is favorable for the change $A \to B$, but that $\Delta H > 0$, which is unfavorable for a change in that direction. Of course, this latter condition is favorable to the reverse process $B \to A$. As an example of such a process, consider ice melting into water, with A being ice and B being water. Ice requires latent heat in order for it to melt, so ΔH is positive. (This is physically reasonable: $\Delta H = \Delta U + \Delta(PV) \approx \Delta U$ for a solid–liquid transition. This approximation is valid because the $\Delta(PV)$ term is generally small compared with the ΔU term, due to the small volume change, and ΔU is positive because the potential energy of the molecules increases when they break free from their bound positions in the solid.) Now, ΔS is also positive, because ice is a more ordered state than water. At low temperatures, the ΔH term wins, so ΔG is positive and the process $A \to B$ does not occur. However, at high temperatures, the $T_0 \Delta S$ term wins, so that $\Delta G < 0$ and the process $A \to B$ goes ahead. This certainly is true for ice at one atmosphere remaining ice at a temperature less than $0°C$ and melting into water at higher temperatures. The stable condition of ice coexisting with water occurs when the ΔH term is exactly balanced by the $T_0 \Delta S$ term at $0°C$ with $\Delta G = 0$. This idea will be discussed again in the larger context of phase transitions in Chapter 10. See also the book by Peters (2010).

7.5.4 Useful Work

Section 7.4 showed that, in a process in which the initial and final temperatures are equal to the temperature of the surroundings and where the heat transferred is between the system and the surroundings only, the maximum work that can be obtained from a system is equal to the decrease in F. Consider a gas in a cylinder expanding through a volume ΔV and by doing so performing some external work W. For example, this could be the lifting of a weight, as illustrated in Figure 7.6(a) or the turning of the shaft of an electrical generator. In performing this work, the gas also does useless work $P_0 \Delta V$ against the surrounding atmosphere at the pressure P_0 and temperature T_0. Another example of useless work occurs in an electrolytic cell where, in addition to the electrical work delivered at the rate $I^2 R$ when a current I is passed through a resistance R, the cell also performs useless work because any gases produced in the electrolytic reaction have to push back the atmosphere. This is illustrated in Figure 7.6(b).

There is an elegant way of subtracting off this useless work from the total work performed by a system that is allowed to expand between a pair of states, both at P_0 and T_0, and where any heat transferred is between the system and the surroundings only. The cell illustrated in Figure 7.6(b) is an example of such

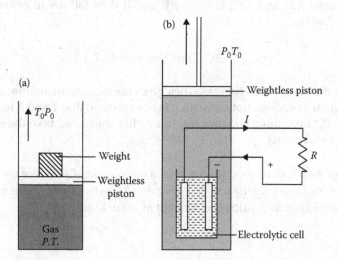

Figure 7.6 Illustrations of useful and useless work. In each case, the system has to expand against the surroundings and, in doing so, has to perform an amount $P_0 \Delta V$ of useless work.

a process; however, the gas-weight system of Figure 7.6(a) must be excluded, because the end point pressures are not P_0, even though the end point temperatures are T_0.

Before proceeding further, consider again the requirement that the end points be at P_0 and T_0. Clearly, in a simple system, this means that there can be no change in volume, and so no volume work can be performed. For a change of volume to occur under these conditions, there must be some additional degree of freedom within the system, as would be allowed, for example, by an internal chemical reaction.

For a process beginning and ending at (P_0, T_0), assuming it to be reversible, Equation 7.31 gives $W = -\Delta F$. Also,

$$W_{useful} = W - P_0\Delta V = -\Delta F - P_0\Delta V \qquad 7.47$$

or

$$W_{useful} = -\Delta(F + P_0\Delta V) = -\Delta(F + PV) \qquad 7.48$$

because the end point pressures are both P_0. As can be seen from the defining Equations 7.21 and 7.37, $G = F + PV$, and it then follows in general from Equation 7.48 that

$$W_{useful} \leq -\Delta G \quad (\text{end points at } P_0, T_0) \qquad 7.49$$

The decrease in the Gibbs function then gives the maximum amount of useful work for such a process. Notice again it is not required that T and P to be fixed at T_0 and P_0 throughout the process, but rather they need take these values only at the end points.

Because of its relation to useful work, G is often called *Gibbs free energy*. The concept of useful work arises again in Section 7.6, when we introduce a rather more general approach using the concept of availability.

7.5.5 F or G?

Sections 7.4 and 7.5 described carefully the properties of the two state functions F and G and the circumstances in which each one is appropriate to use.

In particular, for a system in thermal and mechanical contact with a heat and pressure reservoir, G minimizes at equilibrium. However, if the system is kept at constant volume and is in thermal contact with a reservoir, it is F that minimizes. Unfortunately, F and G frequently appear to be used interchangeably in the literature in a somewhat loose manner, and so the validity of the procedure should be examined.

By definition, $F = U - TS$ and $G = U + PV - TS$. Therefore, the two functions are related by $G = F + PV$. In any change,

$$\Delta G = \Delta F + \Delta(PV)$$

In some cases, the change $\Delta(PV)$ is small compared to $\Delta F = \Delta(U) - \Delta(TS)$. This would occur, for example, in a process involving a metal, where the volume change is usually small. In that case $\Delta G \approx \Delta F$, and little error is incurred in using one function instead of the other. However, in a process involving a gas, the term $\Delta(PV)$ is important because of large volume changes ΔV. In general, one should be very careful to use the correct potential function appropriate to the circumstances.

7.6 AVAILABILITY FUNCTION *A*

The arguments used in the previous sections in deriving the conditions for equilibrium under different conditions can be generalized using the concept of *availability*.

From Section 7.5, the condition for equilibrium for a system in thermal and mechanical contact with a heat and pressure reservoir is that the Gibbs function for the system is a minimum. Further, minimizing G for the system automatically maximizes the entropy of the system and the surroundings. Section 7.4 showed that, for a system in thermal contact with a heat reservoir but kept at constant volume, the equilibrium condition was one of minimum F for the system. Again, minimizing F automatically maximizes the entropy of the system and the surroundings. These results both follow from the concept of availability.

As a starting point, consider a system at P, V, and T that can exchange heat Q with the surroundings, a heat and pressure reservoir at P_0 and T_0, in a process

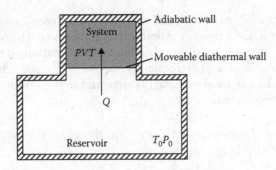

Figure 7.7   A system in contact with a reservoir at T_0 and P_0. Such an arrangement tends to a state of minimum availability.

on the way to equilibrium. In this process, the volume of the system may change by ΔV. This is shown in Figure 7.7. In the general case, there are no restrictions on the state variables of the system; these will be introduced later in the argument.

The reasoning follows the now familiar pattern. Heat Q flows between the reservoir and the system, with an adiabatic wall around the outside so that no heat crosses this boundary. Hence,

$$\Delta S + \Delta S_0 \geq 0 \qquad\qquad 7.50$$

by the principle of increasing entropy for the combined system and reservoir. But with $\Delta S_0 = -Q/T_0$ for the reservoir, the entropy increasing principle becomes $-Q/T_0 + \Delta S \geq 0$ or

$$Q - T_0 \Delta S \leq 0 \qquad\qquad 7.51$$

From the first law $\Delta U = Q - P_0 \Delta V$. Substituting for Q in Equation 7.51 gives

$$\Delta U + P_0 \Delta V - T_0 \Delta S \leq 0 \qquad\qquad 7.52$$

So far this follows the familiar pattern. Now define the *availability A* as

---

Availability function A:

$$A = U + P_0 V - T_0 S \qquad\qquad 7.53$$

Notice that A is a function of the state variables of both the system and the surroundings. With this definition, Equation 7.52 becomes

$$\Delta A \leq 0 \qquad\qquad 7.54$$

This is an entirely general result:

The availability can only decrease, becoming a minimum at equilibrium.

Now, one may introduce the particular restrictions on P, V, and T. In the first special case of a system in thermal and mechanical contact with a heat and pressure reservoir, $P = P_0$ and $T = T_0$ at the start and end of the process (though not necessarily at the intermediate stages). Then,

$$\Delta A = \Delta(U + P_0 V - T_0 S) = \Delta(U + PV + TS) = \Delta G \leq 0 \qquad 7.55$$

with G of the system becoming a minimum at equilibrium. This agrees with the result in Equation 7.45.

Now consider the second special case of a system kept at constant volume and in thermal contact with a heat reservoir

$$\Delta A = \Delta(U + P_0 V - T_0 S) = \Delta(U - T_0 S) = \Delta(U - TS) = \Delta F \leq 0 \qquad 7.56$$

This result follows because V is unchanged and $T = T_0$ at the beginning and end of the process. This matches the earlier result in Equation 7.34.

In a similar way, the equilibrium conditions for a system held under different external conditions from the two cases considered here (which are the most important ones) may be readily obtained using the general result in Equation 7.54 for availability. (See the book by Adkins (1984) for a complete list of all the possibilities.) Further, it is quite clear from the definition of A that these results follow from maximizing the entropy of both the system and the surroundings, whereas in minimizing G or F of the system alone, this point can be all too easily misunderstood.

You can see why the name "availability" is appropriate. The decrease in the availability is equal to the maximum amount of useful work that can be extracted from a system in a given set of surroundings. In other words, it gives the amount of energy *available* for useful work. The proof of this property is quite straightforward and similar to the discussion of G in Section 7.5. There you saw that, in a system performing work W, an amount of useless work $P_0 \Delta V$ was expended against the surroundings and was not available for useful work. In the example of Figure 7.6(b) of the electrolytic cell, this work could not be used as useful work in either Joule heating in the resistor or in driving an external electric motor. The useful work is then

$$W_{useful} = W - P_0 \Delta V \qquad 7.57$$

The first law, applied to the system, is

$$\Delta U = Q - W = Q - \left(W_{useful} + P_0 \Delta V \right) \qquad 7.58$$

Substituting the value of Q given by Equation 7.51,

$$\Delta U + W_{useful} + P_0 \Delta V - T_0 \Delta S \leq 0 \qquad 7.59$$

or

$$W_{useful} + \left(\Delta U + P_0 \Delta V - T_0 \Delta S \right) \leq 0 \qquad 7.60$$

Hence,

$$W_{useful} \leq -\Delta A \qquad 7.61$$

For a given change of state in the system in a given environment, ΔA is fixed (because ΔU, ΔV, ΔS, P_0, and T_0 are all fixed), and the maximum useful work that can be obtained will be when the equality sign in Equation 7.61 holds. Therefore,

$$\left(W_{useful} \right)_{maximum} = -\Delta A \qquad 7.62$$

which proves the proposition.

Of course, useless work is also done against friction and other dissipative forces that have so far been ignored. If the work $W_{friction}$ done against friction is included, Equation 7.61 becomes

$$W_{\text{friction}} + W_{\text{useful}} \leq -\Delta A \qquad\qquad 7.63$$

The effect of the extra term on the left side of Equation 7.63 is to make the inequality in Equation 7.61 even stronger. Thus, the inequality $W_{\text{useful}} \leq -\Delta A$ is entirely general; it is valid even in the presence of dissipative forces.

As an illustration of these rather general ideas (the fact that they are general is the power of the availability approach), consider once again the specific and most important case of a system in thermal and mechanical contact with the surroundings at P_0 and T_0 so that, in any process $\Delta T = \Delta P = 0$ for the system. As usual, this follows because $P = P_0$ and $T = T_0$ at the end points, if not in between. Equation 7.61 becomes

$$W_{\text{useful}} \leq \Delta(U + P_0 V - T_0 S) = -\Delta(U + PV - TS) = -\Delta G$$

with

$$\left(W_{\text{useful}}\right)_{\text{maximum}} = -\Delta G$$

This is the result obtained earlier as Equation 7.49. This should cause no surprise, because both results follow from the same underlying idea: the principle of increasing entropy.

7.7 THE THERMODYNAMIC SQUARE

This chapter contains a large number of equations, variables, and relationships. Fortunately, there are patterns that make it easier to remember the most important relationships. This section introduces a way to organize them.

7.7.1 Thermodynamic Identities

Section 5.5 presented the first thermodynamic identity

$$dU = TdS - PdV \qquad\qquad 5.10$$

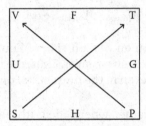

Figure 7.8   The thermodynamic square, which shows relationships between thermodynamic potentials and state functions and yields the four thermodynamic identities.

In this chapter, similar equations were found for differentials of the other thermodynamic potentials

$$dH = TdS + VdP \qquad\qquad 7.11$$

$$dF = -PdV - SdT \qquad\qquad 7.23$$

$$dG = VdP - SdT \qquad\qquad 7.38$$

These four equations (not just the first one) are collectively known as the four thermodynamic identities. A little examination reveals that they are cyclic permutations of one another, and this fact can be used to construct a visual aid called the *thermodynamic square* (Figure 7.8), which makes it easier to remember the relationships.

For example, suppose you want to remember the first thermodynamic identity, containing dU. The first step is to find U on the left side of the square. Then, U is adjacent to S and V, which in turn are connected by lines to P and T. This pattern means that the equation for dU contains the terms PdV and TdS. The arrows pointing toward V and away from S indicate that the correct terms are $-PdV$ and TdS. That is, an arrow pointing toward an adjacent potential indicates a minus sign and an arrow pointing away indicates a plus sign. Similarly, the equation for dH contains the terms $+TdS$ and $+VdP$, and so on for the other two identities.

7.7.2 Maxwell Relations

The four Maxwell relations, which contain P, V, T, and S, are also related in a regular way that allows them to be reconstructed using the same square. The four relations are

$$\left(\frac{\partial T}{\partial V}\right)_S = -\left(\frac{\partial P}{\partial S}\right)_V \; (\text{from } U) \qquad\qquad 7.5$$

$$\left(\frac{\partial T}{\partial P}\right)_S = \left(\frac{\partial V}{\partial S}\right)_P \; (\text{from } H) \qquad\qquad 7.14$$

$$\left(\frac{\partial P}{\partial T}\right)_V = \left(\frac{\partial S}{\partial V}\right)_T \; (\text{from } F) \qquad\qquad 7.26$$

$$\left(\frac{\partial V}{\partial T}\right)_P = -\left(\frac{\partial S}{\partial P}\right)_T \; (\text{from } G) \qquad\qquad 7.41$$

Notice that the quantities P, V, T, and S occupy the four corners of the square. To obtain any relation, start at one corner of the square. For example, start at the upper-right corner (T) and proceed counterclockwise in order T, V, S, which is read as the left side of Equation 7.5. Then, go to the corner not yet used (P) and proceed clockwise in order P, S, V, read as the right side of Equation 7.5. To determine whether or not to include a minus sign in the Maxwell relation, draw in the counterclockwise and clockwise paths ending in arrows, as shown in Figure 7.9. In this case, the arrows outside the square point in opposite directions, up and down, while those inside the square are in the same direction, both up. This difference signals you to include a minus sign, which yields the correct identity in Equation 7.5. Following the same steps, starting from each of the four corners, yields all four Maxwell relations. The plus signs in Equations 7.14 and 7.26 result from the fact that the arrows inside the square match those outside.

The thermodynamic square is just a memory device, not a proof of the thermodynamic identities or Maxwell's relations. The proofs require a careful understanding of this chapter.

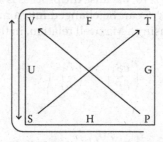

Figure 7.9 Illustration of paths taken around the square that lead to the four Maxwell relations.

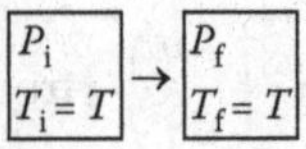

$$\boxed{\begin{array}{c} P_i \\ T_i = T \end{array}} \rightarrow \boxed{\begin{array}{c} P_f \\ T_f = T \end{array}}$$

Figure 7.10   Illustration of the isothermal compression of a block from initial state (P_i, T) to final state (P_f, T).

7.8 EXAMPLE USING A MAXWELL RELATION

Suppose you have a block of metal and compress it reversibly and isothermally, at constant temperature T, from a pressure P_i to a pressure P_f, as illustrated in Figure 7.10. According to the first law, heat will flow out of the block. How much?

Start with the basic equation for heat flow in a reversible process

$$đQ_R = TdS \qquad\qquad 5.4$$

Hence, the problem can be solved by finding dS in terms of known quantities and then integrating. As in the examples given in Sections 2.1.4 and 2.4, one may express S as a function of the state variables for which the changes are given, in this case T and P. That is, $S = S(T, P)$. This is allowed, because the process is reversible, and this is just a way of writing down the equation of state, which holds at each point in the process. Using this procedure,

$$dS = \left(\frac{\partial S}{\partial P}\right)_T dP + \left(\frac{\partial S}{\partial T}\right)_P dT$$

so

$$đQ_R = TdS = T\left(\frac{\partial S}{\partial P}\right)_T dP + T\left(\frac{\partial S}{\partial T}\right)_P dT$$

where the second term is zero, because the process is isothermal. The first partial derivative containing S can be changed into a more familiar expression involving P, V, and T by using a Maxwell relation, in this case Equation 7.41

$$\left(\frac{\partial S}{\partial P}\right)_T = -\left(\frac{\partial V}{\partial T}\right)_P \qquad\qquad 7.41$$

Hence,

$$đQ_R = -T\left(\frac{\partial V}{\partial T}\right)_P dP = -TV\beta dP$$

where $\beta = (1/V)\partial V/\partial T$ is the thermal expansion coefficient from Equation 2.2. Integrating

$$Q = -\int_{P_i}^{P_f} TV\beta\, dP \approx -TV\beta \int_{P_i}^{P_f} dP$$

$$\approx -TV\beta\left(P_f - P_i\right)$$

where it is assumed that there is only a small change of volume in the compression and that the expansion coefficient β is constant. The minus sign indicates that heat flows out, as expected.

A good number of problems in thermodynamics can be solved using this approach.

Problems

7.1   It is a result of statistical mechanics that the internal energy of an ideal gas is

$$U = U(S,V) = \pm Nk_B \left(\frac{N}{V}\right)^{2/3} e^{2S/3Nk_B}$$

where α is a constant and the other symbols have their usual meanings. Show that the equation of state $PV = nRT$ follows from this equation.

7.2   The Helmholtz function of one mole of a certain gas is

$$f = \frac{F}{n} = -\frac{a}{v} - RT\ln(v-b) + j(T)$$

where a and b are constants and j is a function of T only. Derive an expression for the pressure of the gas.

7.3   The table gives the values of some thermodynamic properties of a substance at two different states, both at the same temperature.

	Temp. (°C)	u (kJ/kg)	s (kJ/(K·kg))	P (N/m²)	v (m³/kg)
Initial state	300	2727	6.364	4×10^6	0.0588
Final state	300	2816	8.538	0.05×10^6	5.29

What is the maximum amount of work that can be extracted from 1 kg of this substance in taking it from the initial state to the final state? Select the relevant data from the table. Incidentally, this substance is superheated steam.

7.4   The Gibbs function of one mole of a certain gas is given by

$$g = RT \ln P + A + BP + \frac{CP^2}{2} + \frac{DP^3}{3}$$

where A, B, C, and D are constants. Find the equation of state of the gas.

7.5   Derive the following equations:

a $U = F - T \left(\dfrac{\partial F}{\partial T} \right)_V = -T^2 \dfrac{\partial}{\partial T} \left(\dfrac{F}{T} \right)_V$

b $C_V = -T \left(\dfrac{\partial^2 F}{\partial T^2} \right)_V$

c $H = G - T \left(\dfrac{\partial G}{\partial T} \right)_P = -T^2 \dfrac{\partial}{\partial T} \left(\dfrac{G}{T} \right)_P$

d $C_P = -T \left(\dfrac{\partial^2 G}{\partial T^2} \right)_P$

7.6   Use the thermodynamic identity along with the equality of mixed second partial derivatives

$$\frac{\partial^2 U}{\partial V \partial S} = \frac{\partial^2 U}{\partial S \partial V}$$

to prove the Maxwell relation in Equation 7.5. Use a similar procedure to prove the other three Maxwell relations.

7.7   In the presence of a catalyst, one mole of nitric oxide (NO) decomposes into nitrogen and oxygen. The initial and final temperatures are 25°C and the process occurs at a pressure of one atmosphere. The entropy change is $\Delta s = 76$ J/(K·mol) and the enthalpy change is $\Delta h = -8.20 \times 10^5$ J/mol. What is the change in the Gibbs free energy and what is the heat produced in the decomposition?

7.8   The Haber process for synthesis of ammonia is historically important because it allowed for mass production of fertilizers for agricultural use as well as explosives. The synthesis follows the reaction

$$N_2 + 3H_2 \rightarrow 2NH_3$$

a Use the data given below to find ΔG for this reaction and for one mole of ammonia, assuming standard conditions of atmospheric pressure and $T = 298$ K.

	s (J/(K·mol))	h (kJ/mol)
N_2	192	0
H_2	131	0
NH_3	193	−46.1

(Note that the tabulated values are all for one mole of reactant/product.)

b You should find that $\Delta G < 0$, consistent with a spontaneous reaction. Why is a special process needed to make this work?

7.9 Water boils at $T = 100°C$ at one atmosphere pressure. In the process, the entropy increase is 109 J/K for each mole of water. Find the molar enthalpy increase.

7.10 A gas cools from a temperature T to the temperature T_0 of the surroundings. There is no change between the initial and final volumes, $\Delta V = 0$, but the volume may vary during the process, and so the gas may perform work. This is indicated in Figure 7.11. Show that the maximum amount of work obtainable from the gas is

$$W_{max} = C_V\left(T - T_0\right) + C_V T_0 \ln\left(\frac{T_0}{T}\right)$$

Hints: (a) Consider the argument leading to Equation 7.28. (b) Hence show $W_{max} < -\Delta U + T_0 \Delta S$. (c) Use Equation 5.11.

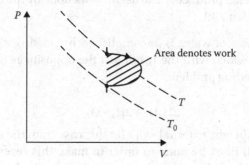

Area denotes work

Figure 7.11

7.11 One mole of an ideal gas expands at the constant temperature T_0 of the surroundings from a pressure P_1 to a pressure P_2. The atmospheric pressure is P_0. (a) By considering the total work done in a reversible expansion and subtracting the useless work, show that the maximum useful work done by the gas is

$$RT_0 = \ln\left(\frac{P_1}{P_2}\right) - P_0 RT_0 \left(1/P_2 - 1/P_1\right)$$

(b) How is this work related to the change in the Gibbs function?

7.12 The hydrogen fuel cell is a device used to power many vehicles (including automobiles) and other machines. It is an important alternative energy source because its emissions are carbon free. In a simple form of fuel cell, hydrogen gas is fed in at one electrode and oxygen at the other. Water is produced according to the reaction

$$2H_2 + O_2 \rightarrow 2H_2O$$

The cell operates at the pressure and temperature (298 K) of the atmosphere. Under those conditions, the molar values of S and H are as follows:

	s (J/(K·mol))	h (kJ/mol)
O_2	205	0
H_2	131	0
H_2O	70	−286

Assume that the cell operates reversibly. (a) Compute the change ΔG in the Gibbs function when one mole of H_2O is produced. (b) Calculate the cell's EMF, which will be equal to the potential difference across the terminals. Hint: Be sure to use values for one mole of H_2O being produced. The useful work done by the cell is then given by Equation 7.49.

7.13 Electrolysis of water is accomplished by passing electrical current through water, with the result just the opposite as in the fuel cell of the preceding problem

$$2H_2O \rightarrow 2H_2 + O_2$$

(a) Explain why external work (in this case from the electrical energy supplied) must be done in order to make this reaction happen. (b) How much work must be done to produce one mole of hydrogen?

7.14 Using the results of the previous two problems, discuss the advantages and disadvantages of the hydrogen fuel cell as an energy source alternative to hydrocarbons.

7.15 (a) For the fuel cell of Problem 7.12, find the work $P\Delta V$ done in producing one mole of water, due to the gases changing into liquid. (b) Find ΔU for this process.

7.16 Show that the equilibrium conditions listed below follow from Equation 7.54 for the availability:

System	Equilibrium Condition
Totally isolated	S a maximum
Thermally isolated, held at constant P	H a minimum
Thermally isolated, held at constant V	U a minimum

7.17 Complete the following steps to prove Equation 7.36. (a) Explain why for a collection of N particles the internal energy is $U = N\bar{E}$. (b) Use the rules of logarithms and the result of Chapter 6 Problem 6.4 to show that

$$U = N\bar{E} = -\frac{N}{Z}\frac{\partial Z}{\partial \beta} = -N\frac{\partial(\ln Z)}{\partial \beta}$$

(c) Recalling that $\beta = 1/k_BT$, show that the result of (b) can be rewritten

$$U = Nk_BT^2\frac{\partial(\ln Z)}{\partial T}$$

(d) Use Equations 7.21 and 7.25 to show that

$$U = F - T\frac{\partial F}{\partial T}$$

(e) Use the result of (d) to show that

$$\frac{\partial}{\partial T}\left(\frac{F}{T}\right) = -\frac{U}{T^2}$$

(f) Combine the results of (c) and (e) to show that $F = -Nk_BT\ln Z$.

7.18 For the paramagnetic material considered in Section 7.4.3, graph the entropy as a function of temperature for a range of temperatures from zero to $5\varepsilon/k_B$. Use the graph to assess the argument in the text regarding low- and high-temperature behavior.

REFERENCES

Adkins, C.J., *Equilibrium Thermodynamics*, third edition, Cambridge University Press, Cambridge, 1984.

Guenault, T., *Statistical Physics*, Springer, New York, 2007.

Peters, A.P.H., *Concise Chemical Thermodynamics*, third edition, CRC Press, Boca Raton, FL, 2010.

Chapter 8: General Thermodynamic Relations

In the first seven chapters of this book, many of the foundations of thermodynamics and statistical physics have been put into place. With these foundations, it is possible to obtain some powerful general results. These results all follow very simply from the basic equations, but it is important to keep the big picture in mind and not get lost in the derivations. To that end, applications of each of the derived results will follow those results.

8.1 DIFFERENCE IN HEAT CAPACITIES, $C_P - C_V$

It was shown in Section 3.5.2 that for an ideal gas

$$C_P = C_V + nR \qquad 3.16$$

This result can be generalized for any system with the state variables, P, V, and T. The starting point comes from either of the two basic relations for the principal heat capacities

$$C_V = T\left(\frac{\partial S}{\partial T}\right)_V \qquad 7.9$$

and

$$C_P = T\left(\frac{\partial S}{\partial T}\right)_P \qquad 7.17$$

The form of Equation 7.9 suggests writing $S = S(T, V)$, so that

$$dS = \left(\frac{\partial S}{\partial T}\right)_V dT + \left(\frac{\partial S}{\partial V}\right)_T dV \qquad 8.1$$

DOI: 10.1201/9781003299479-8

In order to obtain C_P, divide both sides of Equation 8.1 by dT at constant pressure and multiply by T. This gives C_P on the left side and C_V on the right side

$$T\left(\frac{\partial S}{\partial T}\right)_P = T\left(\frac{\partial S}{\partial T}\right)_V + T\left(\frac{\partial S}{\partial V}\right)_T\left(\frac{\partial V}{\partial T}\right)_P \qquad 8.2$$

or

$$C_P = C_V + T\left(\frac{\partial S}{\partial V}\right)_T\left(\frac{\partial V}{\partial T}\right)_P \qquad 8.3$$

> **More formally than the procedure just described, one should write dS = $(\partial S/\partial T)_P\ dT + (\partial S/\partial P)_T\ dP$ and $dV = (\partial V/\partial T)_P\ dT + (\partial V/\partial P)_T\ dP$, and then compare coefficients of dT.**

The aim is to produce a relation between C_P and C_V in terms of quantities that can be measured. In this case, the appropriate quantities are the thermal expansion coefficient β and bulk modulus B, where

$$\beta = \frac{1}{V}\left(\frac{\partial V}{\partial T}\right)_P \qquad 2.2$$

and

$$B = -V\left(\frac{\partial P}{\partial V}\right)_T = \frac{1}{\kappa} \qquad 2.3$$

The second partial derivative $(\partial V/\partial T)$ in the second term on the right of Equation 8.3 is therefore equal to βV. The first partial derivative $\partial S/\partial V$ in that term is not recognizable, but it can be made so by applying a Maxwell relation. The appropriate one is

$$\left(\frac{\partial P}{\partial T}\right)_V = \left(\frac{\partial S}{\partial V}\right)_T \qquad 7.26$$

Hence, Equation 8.3 becomes

$$C_P + C_V + T\left(\frac{\partial P}{\partial T}\right)_V\left(\frac{\partial V}{\partial T}\right)_P \qquad 8.4$$

The first partial derivative in Equation 8.4 has P, V, and T in the wrong order to use K or β directly, but this can be repaired using the cyclical rule of Appendix B

$$\left(\frac{\partial P}{\partial T}\right)_V \left(\frac{\partial V}{\partial P}\right)_T \left(\frac{\partial T}{\partial V}\right)_P = -1$$

or

$$\left(\frac{\partial P}{\partial T}\right)_V = -\left(\frac{\partial P}{\partial V}\right)_T \left(\frac{\partial V}{\partial T}\right)_P \qquad 8.5$$

Hence, Equation 8.4 becomes

$$C_P = C_V - T\left(\frac{\partial V}{\partial T}\right)_P^2 \left(\frac{\partial P}{\partial V}\right)_T$$

Now the results in Equations 2.2 and 2.3 can be substituted to find

$$C_P = C_V + T\beta^2 BV \qquad 8.6$$

which is the final result. As a check, note that the SI units for the last term are

$$(K)\left(K^{-1}\right)^2 \left(N/m^2\right)\left(m^3\right) = N\cdot m/K = J/K$$

which matches the units for heat capacity in the other two terms.

Equation 8.6 is a surprising result, because it relates apparently unconnected parameters: heat capacity, thermal expansion coefficient, and bulk modulus. It is a useful relation because experiments usually measure C_p while theory gives C_V, and so the two can be compared. Also, because the bulk modulus B is positive for all known substances, this means that $C_p > C_V$ in all cases. It is left to the reader (Problem 8.1) to show that Equation 8.6 reduces to Equation 3.16 for an ideal gas.

8.2 EVALUATION OF $(\partial C_V/\partial V)_T$ AND $(\partial C_p/\partial P)_T$

For an ideal gas, internal energy is a function of temperature only: $U = U(T)$. Then,

$$C_V = \left(\frac{\partial U}{\partial T}\right)_V = \frac{dU}{dT}$$

is likewise a function of temperature only. In fact, for an ideal monatomic gas, $U = 3N_A k_B T/2$ per mole, and so $C_V = 3N_A k_B/2 = 3R/2$ is constant. The ideal gas is typical of many systems, in that theory often explains how the internal energy, and hence C_V, varies with T. However, one might also wonder how C_V varies with V; in other words, what is $(\partial C_V/\partial V)_T$?

Before calculating this quantity, we should explain what appears to be a contradiction in terms. If C_V is the heat capacity measured at constant volume, what does it mean to ask for its variation with volume? The answer is that a measurement of C_V involves holding the volume constant and measuring the limiting value of the heat put in divided by the resulting temperature rise as $\delta T \to 0$. If then a different constant value of V is chosen and the measurement for C_V is repeated, the result can be a different value of C_V appropriate to this new value of V. It is in this sense that C_V could depend on V, and so it is sensible to compute the value of $(\partial C_V/\partial V)_T$.

From before

$$C_V = T\left(\frac{\partial S}{\partial T}\right)_V \qquad\qquad 7.9$$

so

$$\left(\frac{\partial C_V}{\partial V}\right)_T = T\left(\frac{\partial}{\partial V}\right)_T\left(\frac{\partial S}{\partial T}\right)_V$$

or

$$\left(\frac{\partial C_V}{\partial V}\right)_T = T\left(\frac{\partial}{\partial T}\right)_V\left(\frac{\partial S}{\partial V}\right)_T \qquad\qquad 8.7$$

because the order of differentiation with respect to V and T may be interchanged in the second-order partial derivative. Applying the Maxwell relation

$$\left(\frac{\partial P}{\partial T}\right)_V = \left(\frac{\partial S}{\partial V}\right)_T \qquad\qquad 7.26$$

to Equation 8.7,

$$\left(\frac{\partial C_V}{\partial V}\right)_T = T\left(\frac{\partial}{\partial T}\right)_V\left(\frac{\partial P}{\partial T}\right)_V$$

or

$$\left(\frac{\partial C_V}{\partial V}\right)_T = T\left(\frac{\partial^2 P}{\partial T^2}\right)_V \qquad\qquad 8.8$$

If one knows the equation of state, Equation 8.8 may be integrated to tell one how C_V varies with V. It is left to the problems (Problem 8.5) to show that C_V for a van der Waals gas is a function of T only.

It is also left as a problem (Problem 8.4) to derive the analogous result

$$\left(\frac{\partial C_P}{\partial P}\right)_T = -T\left(\frac{\partial^2 V}{\partial T^2}\right)_P \qquad\qquad 8.9$$

8.3 ENERGY EQUATION

Internal energy $U = U(T)$ (meaning a function of T only) for the special case of an ideal gas. However, in general U will be a function of volume too, so it should be possible to find an expression for $(\partial U/\partial V)_T$ in terms of P, V, and T.

The basic thermodynamic identity is

$$dU = TdS - PdV \qquad\qquad 7.1$$

so

$$\left(\frac{\partial U}{\partial V}\right)_T = T\left(\frac{\partial S}{\partial V}\right)_T - P \qquad\qquad 8.10$$

Applying the Maxwell relation

$$\left(\frac{\partial P}{\partial T}\right)_V = \left(\frac{\partial S}{\partial V}\right)_T \qquad\qquad 7.26$$

to Equation 8.10 gives

$$\left(\frac{\partial U}{\partial V}\right)_T = T\left(\frac{\partial P}{\partial T}\right)_V - P \quad \text{(energy equation)} \qquad\qquad 8.11$$

Equation 8.11 is a powerful result known as the *energy equation*. If the equation of state is known for the system, then the energy equation gives useful information about the internal energy.

As an example of its application, consider an ideal gas with the equation of state $PV = nRT$. Then,

$$\left(\frac{\partial U}{\partial V}\right)_T = T\frac{nR}{V} - P = 0$$

This result says that U is not an explicit function of V but only of T: $U = U(T)$.

It is straightforward to show why U cannot be an explicit function of P for an ideal gas. Suppose

$$U = f(T, P) \tag{8.12}$$

where f is a function. Then, using the equation of state, one could substitute $P = nRT/V$ in Equation 8.12 to obtain a new function, g, where

$$U = g(T, V) \tag{8.13}$$

which, as was just shown from the energy equation, is untrue. Therefore, Equation 8.12 is false, and $U = U(T)$ only for an ideal gas—the result stated in Section 3.5.

There is a second energy equation that is used less frequently than the first (Equation 8.11). It is left as a problem (Problem 8.6) to show that

$$\left(\frac{\partial U}{\partial P}\right)_T = -\left[T\left(\frac{\partial V}{\partial T}\right)_P + P\left(\frac{\partial V}{\partial P}\right)_T\right] \quad \text{(second energy equation)} \tag{8.14}$$

It is not recommended that students memorize the energy equations. It is sufficient to remember that they exist and the form of the left-hand sides.

8.4 RATIO OF HEAT CAPACITIES, C_P/C_V

Just as there is a useful relation for the difference in the principal heat capacities, so there is another useful expression for their ratio. The result is that

$$\frac{C_P}{C_V} = \frac{\kappa_T}{\kappa_S}$$

where, as in Equation 2.3, κ_T is the usual isothermal compressibility and κ_S is the adiabatic compressibility

$$\kappa_T = -\frac{1}{V}\left(\frac{\partial V}{\partial P}\right)_T \quad \text{and} \quad \kappa_S = -\frac{1}{V}\left(\frac{\partial V}{\partial P}\right)_S$$

Using Equation 7.9 and 7.17,

$$\frac{C_P}{C_V} = \frac{T(\partial S/\partial T)_P}{T(\partial S/\partial T)_V} = \frac{(\partial S/\partial T)_P}{(\partial S/\partial T)_V} \qquad 8.15$$

In this case, S cannot be eliminated from Equation 8.15 by using the Maxwell relations. Indeed, it would be unwise to do so, because S is needed outside the partial derivative to give κ_S. Instead, recast the order in the partial derivatives in Equation 8.15 using the cyclical rule (Appendix B)

$$\left(\frac{\partial S}{\partial T}\right)_P \left(\frac{\partial P}{\partial S}\right)_T \left(\frac{\partial T}{\partial P}\right)_S = -1$$

so

$$\left(\frac{\partial S}{\partial T}\right)_P = -\left(\frac{\partial S}{\partial P}\right)_T \left(\frac{\partial P}{\partial T}\right)_S \qquad 8.16$$

and similarly,

$$\left(\frac{\partial S}{\partial T}\right)_V = -\left(\frac{\partial S}{\partial V}\right)_T \left(\frac{\partial V}{\partial T}\right)_S \qquad 8.17$$

Substituting Equations 8.16 and 8.17 into Equation 8.15,

$$\frac{C_P}{C_V} = \frac{(\partial S/\partial P)_T (\partial P/\partial T)_S}{(\partial S/\partial V)_T (\partial V/\partial T)_S} = \frac{(\partial S/\partial P)_T (\partial V/\partial S)_T}{(\partial V/\partial T)_S (\partial T/\partial P)_S}$$

$$= \frac{(\partial V/\partial P)_T}{(\partial V/\partial P)_S}$$

Using the definitions of the two forms of compressibility from above,

$$\frac{C_P}{C_V} = \frac{\kappa_T}{\kappa_S} \qquad\qquad 8.18$$

At first sight, this may again be a surprising result, relating elasticity parameters to heat capacities. However, it should be realized that the heat capacities are determined under definite conditions of P and V, while the two versions of compressibility are determined under conditions of definite T and S. The Maxwell relations link these particular quantities. Hence, it is not really so surprising that results such as Equations 8.18 and 8.6 exist.

As an application of Equation 8.18, consider a weakly magnetic system, where the infinitesimal work term is given by $B_0\, d\mathcal{M}$ as in Equation 2.13, and where magnetic work is the only work. (That is, the mechanical work $-PdV = 0$.) Then, V has to be replaced by $\mathcal{M}$ and P by $-B_0$. The corresponding relation to Equation 8.18 is

$$\frac{C_{B_0}}{C_{\mathcal{M}}} = \frac{\chi_T}{\chi_S} \qquad\qquad 8.19$$

where the isothermal and adiabatic differential susceptibilities are

$$\chi_T = \frac{\mu_0}{V}\left(\frac{\partial \mathcal{M}}{\partial B_0}\right)_T \quad \text{and} \quad \chi_S = \frac{\mu_0}{V}\left(\frac{\dot{\partial}\mathcal{M}}{\partial B_0}\right)_S \qquad 8.20$$

Thus, by measuring the ratio of the magnetic susceptibilities, which are relatively easy to measure, the ratio of the magnetic specific heats may be determined.

In a paramagnetic salt, the magnetism is due to unpaired effective electron spins. Because of its use as a thermometer below 1 K, the best-known paramagnetic salt is cerium magnesium nitrate, $Ce_2Mg_3(NO_3)_{12}\cdot 24H_2O$, where the

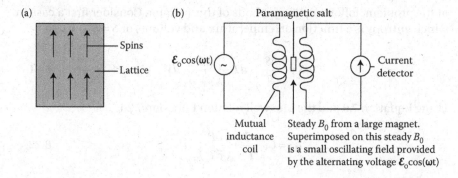

Figure 8.1   Relaxation method of measuring the ratio of magnetic heat capacities.

single unpaired spin is on the cerium ion. One may regard the cerium spins as the system while the magnesium nitrate lattice can be regarded as the surroundings, as shown in Figure 8.1(a). Susceptibilities may be measured using the AC method shown in Figure 8.1(b).

The spins are aligned with a steady magnetic field and, superimposed on this, is a small AC field that flips the spins parallel and antiparallel to the steady field. The current measured by the detector is proportional to the susceptibility χ of the sample.

The spins can flip over in a characteristic time τ, giving up their excess magnetic energy to the lattice. For cerium magnesium nitrate, this time is ~ 0.01 s at $T = 2$ K. At low measuring frequencies, with a period much longer than τ, the spins have plenty of time to exchange their energy of magnetization with the lattice, and so χ_T can be measured. Conversely, χ_S can be measured at high frequencies. From these measurements of χ_T and χ_S, the ratio $C_{B_0}/C_{\mathcal{M}}$ is determined.

> This technique is known as the *relaxation method of measuring heat capacities* and is important in low temperature physics.

8.5   REVISITING THE ENTROPY OF AN IDEAL GAS

Recall that the entropy of an ideal gas was found first in Equation 5.11 and then again using statistical methods in Section 6.2.5. It is useful to take another look at the problem, following the methods of this chapter. Consider first a case in which entropy is a function of temperature and volume, or $S = S(T, V)$, so

$$dS = \left(\frac{\partial S}{\partial T}\right)_V dT + \left(\frac{\partial S}{\partial V}\right)_T dV \qquad 8.21$$

Using Equation 7.9 and the Maxwell relation Equation 7.26,

$$dS = C_V \frac{dT}{T} + \left(\frac{\partial P}{\partial T}\right)_V dV \qquad 8.22$$

Now consider one mole of the gas with the molar heat capacity at a constant volume c_v. Then

$$Pv = RT$$

and

$$\left(\frac{\partial P}{\partial T}\right)_v = \frac{R}{v}$$

Hence, Equation 8.22 becomes

$$ds = c_v \frac{dT}{T} + R \frac{dv}{v}$$

Integrating

$$s = c_v \ln T + R \ln v + s_0 \qquad 5.11$$

where s_0 is a constant. Note that this matches not only Equation 5.11, but also the identical result obtained statistically in Section 6.2.5. For n moles, this becomes

$$S = n\left(c_v \ln T + R \ln v + s_0\right) \qquad 8.23$$

The same procedure may be used to find the entropy as a function of temperature and pressure. Now, $S = S(T, P)$, and so

$$dS = \left(\frac{\partial S}{\partial T}\right)_P dT + \left(\frac{\partial S}{\partial P}\right)_T dP \qquad \text{8.24}$$

By Equation 7.17, $C_P = T(\partial S/\partial T)_P$. Employing this result along with the Maxwell relation in Equation 7.41, Equation 8.24 becomes

$$dS = C_P \frac{dT}{T} - \left(\frac{\partial V}{\partial T}\right)_P dP \qquad \text{8.25}$$

For one mole, the equation of state gives $(\partial V/\partial T)_P = R/P$, so this becomes

$$ds = c_P \frac{dT}{T} - R \frac{dP}{P} \qquad \text{8.26}$$

Integrating Equation 8.26,

$$s = c_P \ln T - R \ln P + s_0 \qquad \text{8.27}$$

for one mole or

$$S = n\left(c_P \ln T - R \ln P + s_0\right) \qquad \text{8.28}$$

for n moles, where s_0 is constant.

8.6 JOULE AND JOULE–KELVIN COEFFICIENTS

The remainder of this chapter is devoted to the study of the Joule coefficient and Joule–Kelvin coefficients. After presenting some basic facts about these concepts, we conclude this chapter by showing how the Joule-Kelvin coefficient is used in refrigeration.

8.6.1 Joule Coefficient for Free Expansion

Recall from Section 2.2.2 the case of free expansion of a gas, as shown in Figure 2.7. It was shown in Section 3.5.1 that no work is done, and if no heat enters the system through the adiabatic walls, then the gas's internal energy is unchanged in the expansion. For an ideal gas, this means that there is no temperature change. Free expansion is an irreversible process. This is seen because the gas does not pass through a series of equilibrium states, and because the entropy of the universe increases in the process (Equation 5.6). However, the end points are equilibrium states, and for this reason thermodynamics can be applied to this process to determine the temperature change.

With no change in internal energy, $U_i = U_f$, where, as usual, the subscripts i and f refer to the initial and final equilibrium points. The temperatures T_i and T_f are uniquely specified by the pairs of variables (U_i, V_i) and (U_f, V_f) and so $\Delta T = T_f - T_i$ is also uniquely specified. Hence, one may imagine a reversible expansion from the state i to the state f and calculate ΔT for the expansion. The ΔT so obtained will be the same as the actual ΔT occurring in the irreversible expansion. The most convenient reversible process connecting the end points is a quasistatic expansion occurring at constant U.

As discussed in Section 2.4, it is now reasonable to write $T = T(V, U)$, so

$$dT = \left(\frac{\partial T}{\partial V}\right)_U dV + \left(\frac{\partial T}{\partial U}\right)_V dU \qquad 8.29$$

where the second term on the right is zero because U is constant in this process. Integrating,

$$\Delta T = \int_{V_i}^{V_f} \left(\frac{\partial T}{\partial V}\right)_U dV \qquad 8.30$$

The partial derivative $(\partial T/\partial V)_U$ is called the *Joule coefficient* μ_J. In order to integrate Equation 8.30, it is necessary to find an expression for μ_J in terms of P, V, and T.

The first thing to recognize about μ_J is that it is difficult to handle as it stands because of the constant U outside the partial derivative. This is in contrast to the usual (simpler) case with U inside a partial derivative, as in $C_V = (\partial U/\partial T)_V$

and in the energy equation, for example. Thus, it is desirable to bring U inside using the cyclical rule (Appendix B)

$$\left(\frac{\partial T}{\partial V}\right)_U \left(\frac{\partial U}{\partial T}\right)_V \left(\frac{\partial V}{\partial U}\right)_T = -1$$

or

$$\left(\frac{\partial T}{\partial V}\right)_U = -\left(\frac{\partial T}{\partial U}\right)_V \left(\frac{\partial U}{\partial V}\right)_T \qquad 8.31$$

This can be used to write the Joule coefficient $\mu_J = (\partial T/\partial V)_U$ as

$$\mu_J = -\frac{1}{C_V}\left(\frac{\partial U}{\partial V}\right)_T \qquad 8.32$$

because $C_V = (\partial U/\partial T)_V$. The partial derivative in Equation 8.32 is given by the energy equation

$$\left(\frac{\partial U}{\partial V}\right)_T = T\left(\frac{\partial P}{\partial T}\right)_V - P \qquad 8.11$$

Therefore, the Joule coefficient becomes

Joule coefficient:

$$\mu_J = \frac{1}{C_V}\left[P - T\left(\frac{\partial P}{\partial T}\right)_V\right] \qquad 8.33$$

The Joule coefficient may be calculated using Equation 8.33 and the equation of state, as shown in the following examples.

8.6.2 Joule Coefficient for an Ideal Gas

The equation of state for one mole of an ideal gas is $Pv = RT$, so

$$\left(\frac{\partial P}{\partial T}\right)_v = \frac{R}{v}$$

Substituting this into Equation 8.33 gives $\mu_J = 0$. That is as expected, because there is no temperature change in the free expansion of an ideal gas.

8.6.3 Joule Coefficient for a Real Gas

A useful modification to $Pv = RT$ for one mole of a real gas is the so-called *virial expansion*:

$$Pv = RT\left(1 + \frac{B_2}{v} + \frac{B_3}{v^2} + \cdots\right) \qquad 8.34$$

where the B_n are the *virial coefficients*, which are temperature dependent and get progressively smaller the higher the order of the term. Therefore, as a first approximation one may keep only the B_2 term and neglect the smaller, higher-order terms. With that approximation, the Joule coefficient becomes (see Problem 8.2)

$$\mu_J = -\frac{1}{c_v}\frac{RT^2}{v^2}\frac{dB_2}{dT} \qquad 8.35$$

The variation of B_2 with temperature is known. For argon, for example, where $dB_2/dT = 2.5 \times 10^{-7}$ m³/(mol·K), μ_J can be calculated from Equation 8.35 to be $\mu_J = -25$ K·mol/m³.

Suppose now that the volume of one mole of argon is doubled at standard temperature and pressure (STP). Equation 8.30 predicts that the expansion will be accompanied by a drop in temperature (because the Joule coefficient is negative). To calculate the expected temperature change, recall that one mole of gas at STP occupies 22.4 L = 0.0224 m³. In order to gain an order of magnitude answer, assume that μ_J is constant for the integration in Equation 8.30. Then,

$$\Delta T \approx \mu_J \Delta v$$

$$\approx \left(-25\,\text{K/m}^3\right) \times \left(0.0224\,\text{m}^3\right) = -0.6\,\text{K}$$

which is a very small effect. Indeed, as noted in Section 3.5.1, Joule was unable to measure this effect for air. This is consistent with the idea that air behaves as a nearly ideal gas under normal conditions. The free expansion of any gas always results in cooling. The physical reason for this was discussed in Section 3.5.1.

8.6.4 Joule–Kelvin Coefficient for the Throttling Process

We conclude this chapter with a discussion of the important Joule–Kelvin effect or throttling process. This is an important effect, because it is widely used in refrigeration and in the liquefaction of gases. The process is illustrated in Figure 3.5. It was shown in Section 3.6 that the process takes place with no change in enthalpy ($H_i = H_f$). Before describing the thermodynamics of this irreversible process, it is useful to describe the effect in more detail.

The most convenient way to represent the effect is on a temperature–pressure plot, as in Figure 8.2. Suppose that the initial equilibrium state before throttling is the point i at (T_i, P_i). With the gas in this initial state, the gas can be throttled to the lower pressure P_{f_1}, where it reaches final temperature T_{f_1} so that the gas ends up in the final equilibrium state (T_{f_1}, P_{f_1}). This state will have the same enthalpy as i. If instead the gas were throttled from the same initial state i to a lower final pressure P_{f_2}, it would then be in the final equilibrium state (T_{f_2}, P_{f_2}), again with the same enthalpy as i.

By repeating this experiment many times, one could obtain a series of points representing the different final equilibrium points, all starting from the same initial equilibrium point and all with the same enthalpy as i. The curve joining them is called an *isenthalp*, as shown in Figure 8.2, because it is a line of constant enthalpy. Remember that an isenthalp is not a curve representing one given throttling process between the equilibrium points at the ends of the

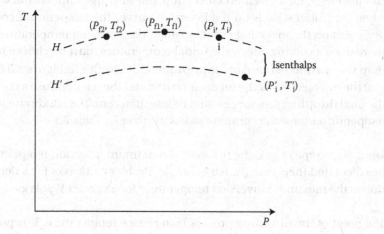

Figure 8.2 Isenthalps of a throttling process for a real (non-ideal) gas.

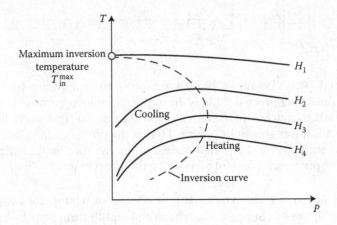

Figure 8.3   Inversion curve for a throttling process for a real (non-ideal) gas.

isenthalp; instead, it is the locus of the end points of different throttling processes, all starting at the same initial state.

If now a second initial point (T_i', P_i') is chosen, a second isenthalp may be drawn. This is shown as the lower curve in Figure 8.2. In a similar way, one can draw a whole series of experimentally determined isenthalps, and the results would appear as in Figure 8.3 for a typical gas. The maxima of the different isenthalps lie on the *inversion curve*. Depending on where the initial and final points are chosen, both heating and cooling effects can be produced. This is in contrast to the free expansion, where there is always cooling. The inversion curve separates the region of heating from the region of cooling. The greatest cooling effect, for a given pressure drop and starting temperature, will occur when the initial state is on the inversion curve. The temperature coordinate of a point on the inversion curve is called the inversion temperature T_{in}. For there to be a cooling effect, the initial temperature must be chosen to be less than the *maximum inversion temperature* T_{in}^{max}. Notice in Figure 8.3 that T_{in}^{max} is at the intersection of the inversion curve and the temperature axis, because the final throttling pressure cannot be less than zero. The maximum inversion temperatures for some common gases are given in Table 8.1.

For a given type of gas, the theoretical maximum inversion temperature can be calculated theoretically. For example, Problem 8.18 asks for a demonstration of the maximum inversion temperature for a van der Waals gas.

The point of the throttling process is to reduce temperature. It is possible to find a general expression for ΔT for a throttling process in terms of P, V, and T.

TABLE 8.1 Maximum Inversion Temperatures for Some Common Gases

Gas	Maximum Inversion Temperature (K) T_{in}^{max}
Argon	723
Nitrogen	621
Hydrogen	205
Helium	51

As for the case of free expansion, the end points at (P_i, H_i) and (P_f, H_f) are equilibrium states, and therefore the temperature change calculated for an imaginary reversible process between them is the same as the temperature change in the actual irreversible throttling process. The most convenient reversible process to choose is a quasistatic expansion from the initial pressure P_i to the final pressure P_f taking place at constant enthalpy. Thus, the temperature is $T = T(P, H)$, so that

$$dT = \left(\frac{\partial T}{\partial P}\right)_H dP + \left(\frac{\partial T}{\partial H}\right)_P dH \qquad 8.36$$

where the second term on the right vanishes for a constant H process. Integrating,

$$\Delta T = \int_{P_i}^{P_f} \left(\frac{\partial T}{\partial P}\right)_H dP \qquad 8.37$$

The partial derivative $(\partial T/\partial P)_H$ is defined to be the *Joule–Kelvin coefficient* μ_{JK}.

In order to perform the integration in Equation 8.37, it is necessary to express μ_{JK} in terms of P, V, and T. The difficulty with

$$\mu_{JK} = \left(\frac{\partial T}{\partial P}\right)_H \qquad 8.38$$

is the constant H outside the partial derivative. It can be brought inside using the cyclical rule and then combined with dT using $C_P = (\partial H/\partial T)_P$. Applying the cyclical rule,

$$\left(\frac{\partial T}{\partial P}\right)_H \left(\frac{\partial H}{\partial T}\right)_P \left(\frac{\partial P}{\partial H}\right)_T = -1$$

or

$$\left(\frac{\partial T}{\partial P}\right)_H = -\left(\frac{\partial T}{\partial H}\right)_P \left(\frac{\partial H}{\partial P}\right)_T \qquad 8.39$$

The first factor on the right of Equation 8.39 is indeed $1/C_P$, while the second factor $(\partial H/\partial P)_T$ is the enthalpy counterpart of the energy equation for U. To solve for this factor, begin with the thermodynamic identity

$$dH = TdS + VdP \qquad\qquad 7.11$$

Taking the partial derivative with respect to P at constant temperature,

$$\left(\frac{\partial H}{\partial P}\right)_T = T\left(\frac{\partial S}{\partial P}\right)_T + V \qquad\qquad 8.40$$

This can be simplified using the Maxwell relation

$$\left(\frac{\partial V}{\partial T}\right)_P = -\left(\frac{\partial S}{\partial P}\right)_T \qquad\qquad 7.41$$

Substituting Equation 7.41 into 8.40,

$$\left(\frac{\partial H}{\partial P}\right)_T = V - T\left(\frac{\partial V}{\partial T}\right)_P \qquad\qquad 8.41$$

Hence, Equation 8.39 becomes

Joule–Kelvin coefficient:

$$\mu_{JK} = \frac{1}{C_P}\left[T\left(\frac{\partial V}{\partial T}\right)_P - V\right] \qquad\qquad 8.42$$

This is the desired result for the Joule–Kelvin coefficient. Now, μ_{JK} can be calculated from the equation of state. Notice that μ_{JK} can be positive or negative (or zero), depending on the relative values of the two terms inside the brackets. Finally, the temperature change in the throttling process may be determined by substituting the value of μ_{JK} given by Equation 8.42 into 8.37, analogous to the process used previously to find ΔT for free expansion, using the Joule coefficient.

It is easily shown that $\mu_{JK} = 0$ for an ideal gas (Problem 8.16), and so there is no temperature change when an ideal gas undergoes a throttling process. For a real gas, μ_{JK} is best calculated from Equation 8.42 by writing the equation of state in a virial form, such as Equation 8.34. All the virial coefficients B_i and their temperature coefficients are tabulated in the reference handbooks such as Kaye and Laby (1995).

The throttling process is of enormous importance, especially in the liquefaction of gases. Gases might also be cooled by allowing them to expand adiabatically in an engine and indeed, it can be shown (see Problem 8.19) that a simple adiabatic expansion produces a bigger temperature drop for a given pressure decrease than a throttling process. Although adiabatic expansion is the more efficient method of cooling, it suffers from the severe disadvantage that moving parts are involved in the expansion engine, and they can seize up at the lowest temperatures, as impurities in the gas freeze out. On the other hand, the throttling process has no moving parts. That is why throttling is always used in the final stages in the liquefaction of nitrogen or helium. The next section explains how the throttling process is used in the classic Linde method of liquefying helium.

> **Another common application of throttling is in cryogenic surgery. Typically, a metal tip is made extremely cold by allowing a gas such as N_2O or CO_2 to throttle through a small opening in the tip. For example, a torn retina can be repaired in this manner, as experienced by this author.**

8.6.5 Linde Liquefaction Process

A schematic representation of the Linde liquefier is given in Figure 8.4. The most difficult gas to liquefy is helium. It has the lowest boiling point at atmospheric pressure, a mere 4.2 K for the most common isotope ^{4}He and 3.2 K for the rare isotope ^{3}He. This discussion involves using the Linde liquefier for helium, but of course it can be used for other gases.

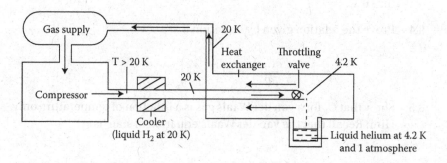

Figure 8.4 Schematic representation of a Linde liquefier.

In order for there to be any cooling using the throttling process, the helium gas has first to be cooled below the maximum inversion temperature of 51 K by passing it through a cooler, which is typically a coiled pipe immersed in a bath of liquid hydrogen at 20 K. The gas then enters the countercurrent heat exchanger at 20 K at high pressure, where it is throttled through a valve. There it undergoes cooling, but not enough to immediately cause liquefaction. The cooled gas passes out through the heat exchanger and, in doing so, cools the incoming gas below 20 K. This gas in its turn expands and cools the next amount of incoming gas even more. Eventually, the temperature on the inlet side of the throttling valve is low enough for liquefaction to occur at the outlet side, and the liquid helium collects at the bottom of the heat exchanger container at 4.2 K and at a pressure of 1 atmosphere.

The compressor drives the helium gas around the circuit and provides the necessary high pressure at the inlet side of the throttling valve. After compression and the consequent heating, the gas is cooled back to 20 K again by the liquid hydrogen cooler so that the helium always enters the heat exchanger at this temperature. It must be understood that the liquefier does not work on the principle that a given mass of gas completes several circuits suffering successive and additive temperature drops until it eventually liquefies.

Problems

8.1   Show that Equation 8.6 reduces to Equation 3.16 for the case of an ideal gas.

8.2   Show that using the approximation in Equation 8.34 with no higher terms than B_2 leads to the Joule coefficient in Equation 8.35.

8.3   Calculate the Joule coefficient for (a) a monatomic gas with $dB_2/dT = 3.6 \times 10^{-7}$ m^3/(mol·K) and (b) a diatomic gas with $dB_2/dT = 2.9 \times 10^{-7}$ m^3/(mol·K). In both cases, assume a pressure of 101.5 kPa at temperature 0°C.

8.4   Prove the relation given by Equation 8.9

$$\left(\frac{\partial C_P}{\partial P}\right)_T = -T\left(\frac{\partial^2 V}{\partial T^2}\right)_P$$

8.5   Show that C_V for a van der Waals gas is a function of temperature only. Hint: Recall that the van der Waals equation of state is

$$\left(P+\frac{a}{v^2}\right)(V-nb)=nRT \qquad\qquad 3.18$$

where a and b are constants and $\upsilon = V/n$ is the molar volume.

8.6 Derive the second energy equation

$$\left(\frac{\partial U}{\partial P}\right)_T = -\left[T\left(\frac{\partial V}{\partial T}\right)_P + P\left(\frac{\partial V}{\partial P}\right)_T\right].$$

8.7 Consider n moles of a van der Waals gas. Show that $(\partial U/\partial V)_T = n^2\, a/V^2$. Hence show that the internal energy is

$$U = \int_0^T C_V dT - \frac{an^2}{V} + U_0$$

where U_0 is a constant. Hint: Express $U = U(T, V)$.

8.8 As in the previous problem, consider n moles of a van der Waals gas. (a) Show that

$$S = \int_0^T \frac{C_V}{T} dT + nR\ln(V - nb) + S_0$$

where S_0 is a constant. Hint: Use $dS = 1/T(dU + PdV)$. (b) Show that the equation for a reversible adiabatic process is

$$T(V - nb)^{nR/C_V} = \text{a constant}$$

if C_V is assumed to be independent of T.

8.9 Show that the difference between the isothermal and the adiabatic compressibility is

$$\kappa_T - \kappa_S = T\frac{V\beta^2}{C_P}.$$

8.10 For each of the following processes, state whether the process is reversible or irreversible, and state which of the quantities S, H, U, F, and G are unchanged: (a) an isothermal quasistatic expansion of an ideal gas in a cylinder fitted with a frictionless piston; (b) same as (a), but for a non-ideal gas; (c) a quasistatic adiabatic expansion of a gas in a cylinder fitted with a frictionless piston; (d) an adiabatic expansion of an ideal gas into a vacuum (a free expansion); and (e) a throttling process of a gas through a porous plug (the Joule–Kelvin effect).

· 8.11　The Carnot cycle takes a particularly simple rectangular form on an ST plot (see Figure 5.11). An SH plot is also useful in engineering. Show that, for an ideal gas as the working substance, a Carnot cycle is again rectangular in this representation. Hint: First derive the general thermodynamic relations

$$\left(\frac{\partial H}{\partial S}\right)_T = T - V\left(\frac{\partial T}{\partial V}\right)_P$$

$$\left(\frac{\partial H}{\partial T}\right)_S = C_P \frac{V}{T}\left(\frac{\partial T}{\partial V}\right)_P$$

8.12　Derive the so-called TdS equations

$$TdS = C_V dT + T\left(\frac{\partial P}{\partial T}\right)_V dV$$

$$TdS = C_P dT - T\left(\frac{\partial V}{\partial T}\right)_P dP$$

$$TdS = C_V\left(\frac{\partial T}{\partial P}\right)_V dP + C_P\left(\frac{\partial T}{\partial V}\right)_P dV$$

8.13　A block of metal of volume V is subjected to an isothermal reversible increase in pressure from P_1 to P_2 at the temperature T. (a) Show that the heat given out by the metal is $TV\beta(P_2 - P_1)$. (b) Show that the work done on the metal is $V\left(P_2^2 - P_1^2\right)/2B$. (c) By using the first law, calculate the change in U. (d) Obtain the same result as in (c) by writing $U = U(T, P)$ so

$$dU = \left(\frac{\partial U}{\partial T}\right)_P dT + \left(\frac{\partial U}{\partial P}\right)_T dP$$

You may assume that β, B, and V are approximately constant during the compression.

8.14　A block of metal is subjected to an adiabatic and reversible increase of pressure from P_1 to P_2. Show that the initial and final temperatures T_1 and T_2 are related as

$$\ln\left(\frac{T_2}{T_1}\right) = \frac{V\beta}{C_P}(P_2 - P_1)$$

You may assume that the volume of the block stays approximately constant during the compression.

8.15 Assuming that helium obeys the van der Waals equation of state, determine the change in temperature when one kilomole of helium gas, initially at 20°C and with a volume of 0.12 m³, undergoes a free expansion to a final pressure of one atmosphere. You should first show that

$$\left(\frac{\partial T}{\partial V}\right)_U = -\frac{a}{C_V}\left(\frac{n}{V}\right)^2$$

Use the following constants: $a = 3.44 \times 10^3$ J·m³/(kmol)²; $b = 0.0234$ m³/kmol; $c_V/R = 1.506$. Hint: You may approximate. First show that $P_2 \gg P_1$. Then, you may take $V_2 \gg V_1$.

8.16 Show that the Joule–Kelvin coefficient is zero for an ideal gas.

8.17 One kilomole of an ideal gas undergoes a throttling process from $P_1 = 4.0$ atm to $P_2 = 1.0$ atm. The initial temperature of the gas is 50°C. (a) What is the temperature change? (b) How much work must be done on the gas to take it reversibly between the initial and final states? (c) What is the entropy change of the gas? Hint: You should calculate the entropy change in two ways: (i) imagine a reversible process in which the gas is taken isothermally between the initial and final states and (ii) apply the general approach of writing $S = S(T, H)$ and then imagining a reversible process in which the pressure is changed at constant H. You will need to show that $(\partial S/\partial P)_H = -V/T$ in general, from $dH = TdS + VdP$, and this reduces to $-nR/P$ for an ideal gas.

8.18 Show that the temperature and volume of the points (T_{in}, V_{in}) on the inversion curve for a van der Waals gas undergoing a Joule–Kelvin expansion are related as

$$T_{in} = 2a\left(V_{in} - nb\right)^2 \left(RbV_{in}^2\right)^{-1}$$

assuming that, at the maximum inversion temperature, $V_{in} \gg nb$, show that $T_{in}^{max} \approx 2a/Rb$.

8.19   Equation 8.42 gives the Joule–Kelvin coefficient

$$\left(\frac{\partial T}{\partial P}\right)_H = \frac{1}{C_P}\left[T\left(\frac{\partial V}{\partial T}\right)_P - V\right]$$

so that the cooling in a throttling process with a pressure change from P_1 to P_2 is

$$\Delta T = \int_{P_1}^{P_2}\left(\frac{\partial T}{\partial P}\right)_H dP$$

In a similar way, show that the cooling in an adiabatic reversible expansion from a pressure P_1 to P_2 is

$$\Delta T = \int_{P_1}^{P_2}\left(\frac{\partial T}{\partial P}\right)_S dP$$

where

$$\left(\frac{\partial T}{\partial P}\right)_S = \frac{T}{C_P}\left(\frac{\partial V}{\partial T}\right)_P$$

Hence, show that, for a given pressure change, the adiabatic expansion produces more cooling than a throttling process. Hint: Consider the difference in the integrands $(\partial T/\partial P)_S - (\partial T/\partial P)_H$ and show that this is positive.

8.20   A sample of nitrogen is initially at $P = 10$ bar and $T = 100$ K, where it has enthalpy -72.0 kJ/kg. It is then throttled to $P = 1.0$ bar and temperature $T = 77.3$ K, where it is in a mixed gas/liquid phase. At that point, the enthalpy of the gas phase is $+88.0$ kJ/kg, and the enthalpy of the liquid is -126.4 kJ/kg. What fraction of the nitrogen becomes liquid after throttling?

REFERENCE

Kaye, G.W.C. and Laby, T.H., Tables *of Physical and Chemical Constants*, Longman, New York, 1995.

Chapter 9: Magnetic Systems

So far in this book, attention has been focused on systems in which the state variables are P, V, and T, which are related by an equation of state so that only two of them are independent. This chapter concerns magnetic systems, which are described by different sets of state variables. In addition to highlighting the practical importance of magnetic systems, this presentation serves to illustrate the wide applicability of the methods of thermodynamics. In Chapter 13, other systems that differ in nature from the familiar fluid PVT systems will be considered.

9.1 THERMODYNAMICS OF MAGNETIC MATERIALS

Unfortunately, the study of electromagnetism is sometimes confused by the use of different systems of units and differing definitions and nomenclature for important quantities. This section begins with a brief review of some fundamentals of electromagnetism. As in the rest of this book, the SI system of units will be used here consistently. The textbook by Griffiths (2017) is recommended as an excellent resource that has become a standard in undergraduate physics for the study of electromagnetism.

9.1.1 Some Fundamentals of Electromagnetism

Electromagnetism is a reciprocal effect, in that magnetic fields are created by subatomic magnetic moments or by moving electric charges, and in turn magnetic moments and moving charges experience forces when placed in a magnetic field. An electric charge Z traveling with velocity v in a magnetic field B experiences a force given by the vector cross product $F = Zv \times B$, a relation often called the *Lorentz force*. This equation deserves elaboration on two small points. First, the symbol Z is used for electric charge here (as introduced

DOI: 10.1201/9781003299479-9

in Section 2.3.3), rather than q or Q as is typical in the study of electromagnetism, so that it will not be confused with heat. Second, the full Lorentz force includes the influence of electric fields E and is thus $F = ZE + Zv \times B$. However, because this chapter is concerned only with magnetism and not electric fields, the first term is omitted for simplicity.

A magnetic material will have a net magnetization M, which is the magnetic moment per unit volume, with SI units A/m. It is related to the magnetic field B by the equation

$$B = \mu_0 \left(H + M \right)$$ 9.1

where H is sometimes called the *auxiliary field* and also has units A/m. H is a convenient quantity in electrodynamics, because Ampere's law can be written in differential form as

$$\nabla \times H = J_f$$

or in integral form as

$$\oint H \cdot d\ell = I_f$$

where the subscripts f refer to the free currents that generate the magnetic field. Thus, I_f is the free current and J_f is the free current density.

Because the Lorentz force law and Maxwell's equations involve vector cross and dot products, the study of electromagnetism is highly dependent on vector calculus. However, for the thermodynamic relations that will be developed here, it will be sufficient (and much easier) to consider the scalar versions of equations, such as writing Equation 9.1: $B = \mu_0(H + M)$. This is a valid representation for linear materials, where the three vectors in Equation 9.1 all lie along the same axis at any point in space. Therefore, only the scalar versions of this equation and those that follow from it will appear in the next sections.

9.1.2 Magnetic Materials and Susceptibility

When studying magnetic materials, it is useful to think of the magnetic field B as composed of two parts: an amount $B_0 = \mu_0 H$ that would be present in free space in the absence of the material, and the contribution $\mu_0 M$ from the material arising from the net circulating currents in the elementary atomic magnets. That is,

$$B = B_0 + \mu_0 M \qquad 9.2$$

In most situations of importance, the density of magnetic moments is fairly uniform throughout the sample. With that condition, the total magnetic moment $\mathscr{M}$ of the sample of volume V is

$$\mathscr{M} = MV \qquad 9.3$$

For many materials, there is a unique dependence of $\mathscr{M}$ on T and B_0, so that

$$\mathscr{M} = \mathscr{M}(B_0, T) \qquad 9.4$$

Because of hysteresis, such a relation does not hold for ferromagnetic materials, and so such materials are excluded from this discussion. It is further found that the magnetization M is proportional to B_0 for many materials.

The measured or bulk *magnetic susceptibility* χ_m in defined terms of the applied fields H and B_0 as

$$\chi_m = \frac{M}{H} = \frac{\mu_0 M}{B_0} = \frac{\mu_0 \mathscr{M}}{V B_0} \qquad 9.5$$

For such linear materials, Equations 9.1 and 9.2 become

$$B = \mu_0 (1 + \chi_m) H = (1 + \chi_m) B_0 \qquad 9.6$$

The discussion that follows will be restricted to the special case of magnetically weak materials where $\chi_m \ll 1$, so that

$$B \approx B_0 \qquad 9.7$$

Unfortunately, the literature reports measured magnetic susceptibilities in different systems of units. As seen in the discussion above (particularly Equation 9.6), χ_m is dimensionless in the SI system, so this is the way susceptibility is presented in Table 9.1. In the CGS system, the susceptibility as defined in Equations 9.5 and 9.6 differs from SI by a factor of 4π. (The CGS value is *smaller* than the SI value by that factor.) An alternate system used is *mass susceptibility*, defined as χ_m divided by the material's mass density, leaving units of m³/kg in SI or cm³/g in CGS. Yet another system is *molar susceptibility*,

TABLE 9.1   Magnetic Susceptibilities of Selected Materials under Standard Temperature and Pressure

	Magnetic Susceptibility χ_m
Paramagnetic materials	
Oxygen (O_2)	1.9×10^{-6}
Sodium	8.4×10^{-6}
Aluminum	2.1×10^{-5}
Gadolinium sulfate $Gd_2(SO_4)_3 \cdot 8H_2O$	3.5×10^{-3}
Diamagnetic materials	
Helium	-9.9×10^{-10}
Nitrogen (N_2)	-5.4×10^{-9}
Water	-9.0×10^{-6}
Diamond	-2.2×10^{-5}

Note: Values given are dimensionless in the SI system.

defined as the mass susceptibility multiplied by the molar mass, which yields units of m³/mol in SI or cm³/mol in CGS.

Notice in Table 9.1 that paramagnetic materials have positive magnetic susceptibility, but diamagnetic materials have negative susceptibility. This indicates that the magnetic dipoles in a paramagnet align with the applied magnetic field, while dipoles in the diamagnetic material align opposite the field, though weakly.

The SI value of χ_m for the paramagnetic salts (which will be of interest in the subsequent discussion) is only approximately 10^{-2} or 10^{-3}, while diamagnetic materials have $\chi_m \sim -10^{-6}$. Thus, the approximation in Equation 9.7 is justified.

EXAMPLE 9.1

From Table 9.1, the magnetic susceptibility of aluminum at standard temperature and pressure is $\chi_m = 2.1 \times 10^{-5}$, given in SI (dimensionless) units. Convert this value to the other primary systems used: CGS, mass susceptibility, and molar susceptibility.

Solution: The dimensionless GCS value is a factor of 4π smaller, or

$$\chi(\text{CGS}) = \frac{2.1 \times 10^{-5}}{4\pi} = 1.7 \times 10^{-6}$$

To compute the other versions, we look up the standard density of aluminum (2700 kg/m^3) and its molar mass ($27.0 \text{ g/mol} = 0.027 \text{ kg/mol}$). The mass susceptibility is

$$\chi_{\text{mass}} = \chi_m / \rho = \frac{2.1 \times 10^{-5}}{2700 \text{ kg/m}^3} = 7.8 \times 10^{-9} \text{ m}^3/\text{kg}$$

Molar susceptibility is

$$\chi_{\text{molar}} = \chi_{\text{mass}} M = \left(7.8 \times 10^{-9} \text{ m}^3/\text{kg}\right)\left(0.027 \text{ kg/mol}\right) = 2.1 \times 10^{-10} \text{ m}^3/\text{mol}$$

These all agree with standard tabulated values.

9.1.3 Curie Law

Practical interest generally focuses on paramagnetic materials, many of which are found to obey the Curie law

$$\chi_m = \frac{C}{T} \quad (\text{Curie Law}) \qquad\qquad 9.8$$

where C is a constant for a given paramagnetic material and is called the *Curie constant*. The Curie law is an empirical rule discovered by Pierre Curie in 1895, but it is straightforward (Problem 9.3) to derive the law from the statistical properties of paramagnetic materials. Strictly speaking, the Curie law is only valid in the high temperature limit. As a practical matter, however, "high temperature" in this context often means any temperature more than a few degrees above absolute zero (see Problem 9.4). The Curie law relates $\mathcal{M}$, B_0, and T. For the purpose of this discussion, the Curie law may be taken as the equation of state for the magnetic system.

For samples that are not magnetically weak and where the interaction between the magnetic ions is important, the Curie–Weiss law

$$\chi_m = \frac{C}{T - T_0} \qquad\qquad 9.9$$

is a more general modification of the Curie law. T_0 is the Curie–Weiss constant, which is usually only a fraction of a degree for the paramagnetic salts that will be considered here. Thus, the Curie law (Equation 9.8) is a good enough approximation for our purposes.

9.1.4 Thermodynamic Relations for Magnetic Materials

As shown in Appendix C, the appropriate form for the infinitesimal work done on a magnetic system is

$$dW = B_0 d\mathcal{M} \qquad\qquad 9.10$$

Thus, the infinitesimal form of the first law is

$$dU = dQ - PdV + B_0 d\mathcal{M} \qquad\qquad 9.11$$

Under normal conditions, there is no change in pressure, and only the applied magnetic field B_0 is varied. The change in volume of a magnetic system upon the application of an external magnetic field is known as *magnetostriction*. This effect is always small and is certainly negligible for the paramagnetic salts that will be considered. Thus in practice the $-PdV$ term may be ignored, compared with the $B_0 d\mathcal{M}$ term in Equation 9.11. Then the first law becomes

$$dQ = dU - B_0 d\mathcal{M} \qquad\qquad 9.12$$

The thermodynamic treatment of a magnetic system follows that for the familiar PVT system, with P replaced by $-B_0$ and V by $\mathcal{M}$. Making these replacements in the thermodynamic square (Figures 7.8 and 7.9) allows one to reproduce the thermodynamic identities and Maxwell relations. The extra minus sign on B_0 should not cause difficulty, because a constant $-B_0$ outside a partial derivative is the same as a constant $+B_0$. The reader should try to produce, for example, the Maxwell relation

$$\left(\frac{\partial S}{\partial B_0} \right)_T = \left(\frac{\partial \mathcal{M}}{\partial T} \right)_{B_0}$$

both from the square and also using the full analysis involving the Gibbs function G.

9.2 MAGNETIC COOLING

One of the exciting frontiers of physics is the temperature region close to absolute zero. Although the technique of magnetic cooling was first used in the 1930s, it is still a standard technique used to achieve subkelvin temperatures (i.e., <1 K). The recent introduction of the ^{3}He–^{4}He dilution refrigerator has meant that temperatures as low as 4×10^{-3} K can be maintained, but temperatures significantly lower than this can be produced using the technique of magnetic cooling. It is possible to produce temperatures as low as 10^{-6} K using this method, in which the elementary magnetic dipoles comprising the magnetic system are the nuclear spins of copper. However, we shall discuss the application of this technique in cooling a set of electron spins in a paramagnetic salt.

> The lowest temperature records in recent years have been achieved by laser cooling, as discussed in Chapter 6.

9.2.1 Theory of Magnetic Cooling

Before presenting a detailed thermodynamic analysis, it is first necessary to understand the principle of the method. A preliminary step is to immerse the sample in a bath of liquid helium at 4.2 K, which is helium's boiling point at $P = 1$ atm. In Chapter 10 it is shown that the liquid–vapor transition occurs at lower temperatures as the pressure is reduced. Thus, the temperature can be reduced to about 1 K by pumping on the helium vapor above the bath. Near 1 K is a practical limit to cooling by this method, as further reduction in pressure leads to more rapid boiling, until the maximum pumping rate is reached for a given mechanical pump.

It is at this point that the process of magnetic cooling begins. Figure 9.1 shows the different stages used in the technique for cooling a paramagnetic salt. The salt is suspended by fine cotton threads in the middle of a chamber immersed in a bath of liquid helium at 1 K. Surrounding the salt is exchange gas, again helium, which may be pumped away so that the salt may be thermally isolated from the surrounding helium bath. The sequence of operations is as follows.

1. The salt is magnetized with the application of a large magnetic field B_0 of the order of several tesla, provided by an ordinary electromagnet or a superconducting magnet; the latter can produce fields as large as 10 T. Because of the presence of the exchange gas, the magnetization is isothermal. The heat of magnetization is conducted away to the helium bath by the exchange gas.

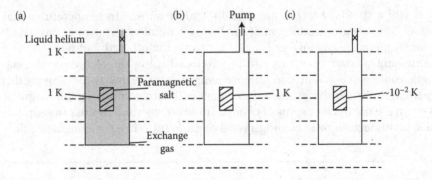

Figure 9.1 Different stages used in adiabatic demagnetization.

2. The exchange gas is pumped away, so that the salt is thermally isolated.

3. The applied magnetic field is slowly reduced to zero so that the salt is always in a state of thermodynamic equilibrium and the demagnetization proceeds reversibly. Because the process is also adiabatic, the demagnetization is thus *isentropic*. The temperature then falls dramatically. For a starting temperature of 1 K, a demagnetization temperature of close to 0.01 K (after many cycles) is typical.

Because of the steps outlined above, this method of magnetic cooling is often referred to as *adiabatic demagnetization*.

To understand the physical reason for this effect, consider the entropy-temperature curves shown in Figure 9.2(a). The upper curve shows the entropy of the salt in the absence of an applied magnetic field. At temperatures of the order of 1 K, only the g-fold degenerate ground energy level of the salt is occupied. The degeneracy g is usually a small number such as 2 or 3. This is illustrated in Figure 9.2(b). Each magnetic atom has g possible ways of achieving the lowest energy, and so the number of different ways of arranging the N atoms is $\Omega = g^N$. The entropy of the magnetic system is then, using Equation 6.3,

$$S = k_B \ln g^N$$

or

$$S = k_B N \ln g \qquad\qquad 9.13$$

In fact, the weak magnetic coupling between the neighboring electron spins splits the states of the ground level, so that they are not actually degenerate. This splitting is so small that at 1 K all these states are thermally occupied, and Equation 9.13 holds. At lower temperatures, however, only the actual ground state is occupied with $\Omega = 1$, and the entropy falls to zero as shown in the figure.

The lower curve of Figure 9.2(a) shows the entropy in an externally applied magnetic field B_0. The application of such a field splits the ground level, as shown in Figure 9.2(c). The degeneracy is then removed, so the entropy is reduced. An alternative viewpoint is to note that the application of B_0 aligns the dipoles, thus imposing more order on the spin system and reducing the entropy. It is a consequence of the third law of thermodynamics (Chapter 12) that the two entropy curves meet at absolute zero.

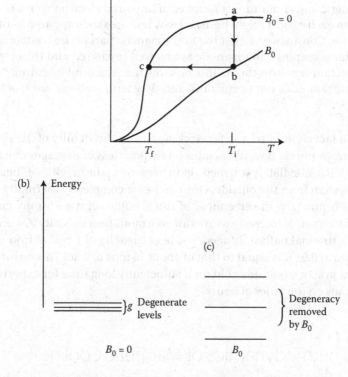

Figure 9.2   (a) Entropy as a function of temperature for a paramagnetic salt in an applied magnetic field B_0 and in zero field. The process ab represents an isothermal magnetization, and the process bc represents an adiabatic demagnetization to the temperature T_f. (b) and (c) The g-fold degeneracy of the energy levels is removed upon the application of B_0.

The physical reason for the drop in temperature is now evident. The isothermal magnetization is represented by the path ab in Figure 9.2(a). The isentropic adiabatic demagnetization is represented by the path bc. As long as the demagnetization process takes place in the region of the shoulder of the entropy curve, significant cooling occurs. To achieve temperatures close to the absolute zero, the magnetic salt should be chosen so that this shoulder is at very low temperatures.

Some typical paramagnetic salts used are

1.   *Iron* ammonium alum.

2.   *Gadolinium* sulfate.

3.   *Cerium* magnesium nitrate.

The magnetic ion, containing a number of unpaired electron spins, is shown in italics in each case. It might be asked why one uses such apparently obscure compounds. The answer is that the non-magnetic part of the compound acts as a dilutant, keeping the magnetic atoms well separated and thus reducing the interaction between them. This ensures that the drop in entropy due to this interaction does not occur until very low temperatures are reached, as required.

There is in fact another reason for working within the vicinity of the shoulder of the entropy curve. It would be absolutely pointless to demagnetize a magnetic salt if it immediately warmed again because of the inevitable "heat leak" into the system from the outside. One can never completely thermally isolate the salt. Fortunately, the steepness of the shoulder of the entropy curve in exactly the region of interest ensures an enormous heat capacity ($C = T \, dS/dT$) to act as a thermal ballast. Indeed, the heat capacity of 1 cm^3 of iron ammonium alum at 0.01 K is equal to that of about 16 tons of lead! This ensures that the salt in practice remains cold for a sufficiently long time for experimental use, perhaps on the order of hours.

9.2.2 Thermodynamics of Magnetic Cooling

It is straightforward to calculate the cooling produced in the adiabatic demagnetization bc shown in Figure 9.2(a). The calculation of the heat produced in the isothermal magnetization ab is left as Problem 9.8.

In the entire process, there are changes in the state functions B_0 and S, so it is possible to write $T = T(B_0, S)$. Therefore,

$$dT = \left(\frac{\partial T}{\partial B_0}\right)_S dB_0 + \left(\frac{\partial T}{\partial S}\right)_{B_0} dS$$

or

$$dT = \left(\frac{\partial T}{\partial B_0}\right)_S dB_0 \qquad 9.14$$

where the second term vanishes because dS is zero for bc.

It would be advantageous to bring the S inside the partial derivative, so that it can be coupled with ∂T to produce heat capacity. This can be done using the cyclical rule (Appendix B)

$$\left(\frac{\partial T}{\partial B_0}\right)_S \left(\frac{\partial S}{\partial T}\right)_{B_0} \left(\frac{\partial B_0}{\partial S}\right)_T = -1 \qquad 9.15$$

or

$$\left(\frac{\partial T}{\partial B_0}\right)_S = -\left(\frac{\partial T}{\partial S}\right)_{B_0} \left(\frac{\partial S}{\partial B_0}\right)_T \qquad 9.16$$

Using $C_{B_0} = T(\partial S/\partial T)_{B_0}$ and the Maxwell relation $(\partial \mathcal{M}/\partial T)_{B_0} = (\partial S/\partial B_0)_T$, this becomes

$$\left(\frac{\partial T}{\partial B_0}\right)_S = -\frac{T}{C_{B_0}}\left(\frac{\partial \mathcal{M}}{\partial T}\right)_{B_0} \qquad 9.17$$

From Section 9.1.2, the magnetic susceptibility is

$$\chi_m = \frac{\mu_0 M}{B_0} = \frac{\mu_0 \mathcal{M}}{VB_0} \qquad 9.5$$

Differentiating with respect to T at constant B_0,

$$\frac{VB_0}{\mu_0}\left(\frac{\partial \chi_m}{\partial T}\right)_{B_0} = \left(\frac{\partial \mathcal{M}}{\partial T}\right)_{B_0} \qquad 9.18$$

Substituting Equation 9.18 into 9.17,

$$\left(\frac{\partial T}{\partial B_0}\right)_S = -\frac{TVB_0}{C_{B_0}\mu_0}\left(\frac{\partial \chi_m}{\partial T}\right)_{B_0} \qquad 9.19$$

Note that the last factor here is given by the Curie law (Equation 9.8) as $(\partial \chi_m / \partial T)_{B_0} = -C/T^2$.

The next step is to substitute Equation 9.19 into 9.14 and integrate with respect to B_0 and T to obtain the temperature fall. Before integrating, however, it is necessary to find the B_0 dependence of $C_{B_0}(T, B_0)$. It might first appear strange to say that the heat capacity at constant B_0 depends on B_0. What this really means is that C_{B_0} takes on different values when measured at different but steady values of B_0.

In exact analogy to Equation 8.9,

$$\left(\frac{\partial C_{B_0}}{\partial B_0}\right)_T = T\left(\frac{\partial^2 \mathcal{M}}{\partial T^2}\right)_{B_0} \qquad 9.20$$

or

$$\left(\frac{\partial C_{B_0}}{\partial B_0}\right)_T = \frac{TVB_0}{\mu_0}\left(\frac{\partial^2 \chi_m}{\partial T^2}\right)_{B_0} \qquad 9.21$$

using Equation 9.5. As well as obeying the Curie law, paramagnetic salts usually have a so-called *Schottky temperature dependence* of the heat capacity C_{B_0} in zero magnetic field, given by

$$C_{B_0}(T,0) = \frac{Vb}{T^2} \qquad 9.22$$

where b is a constant. Using Equation 9.8 in 9.21,

$$\left(\frac{\partial C_{B_0}}{\partial B_0}\right)_T = \frac{2VB_0C}{\mu_0 T^2} \qquad 9.23$$

integrating at constant temperature,

$$C_{B_0}(T,B_0) = \frac{VCB_0^2}{\mu_0 T^2} + C_{B_0}(T,0)$$

The integration constant $C_{B_0}(T,0)$ is given by Equation 9.22, so

$$C_{B_0}(T,B_0) = \frac{V}{T^2}\left(b + \frac{CB_0^2}{\mu_0}\right) \qquad 9.24$$

Substituting Equation 9.19 into 9.14,

$$dT = \left(\frac{\partial T}{\partial B_0}\right)_S dB_0 = -\frac{TVB_0}{C_{B_0}\mu_0}\left(\frac{\partial \chi_m}{\partial T}\right)_{B_0} dB_0$$

Using the expression for C_{B_0} given in Equation 9.24 along with $(\partial \chi_m/\partial T)_{B_0} = -C/T^2$,

$$dT = \frac{TVB_0}{\mu_0} \frac{1}{(V/T^2)(b+(CB_0^2/\mu_0))}\left(\frac{C}{T^2}\right) dB_0 \qquad 9.25$$

Rearranging,

$$\frac{1}{T} dT = \frac{CB_0}{\mu_0\left(b+(CB_0^2/\mu_0)\right)} dB_0 \qquad 9.26$$

This is the desired form, which can be integrated from initial state i to final state f to find

$$\ln T\big|_i^f = \frac{1}{2}\ln\left(b+\frac{CB_0^2}{\mu_0}\right)\bigg|_i^f \qquad 9.27$$

or

$$\frac{T_f}{T_i} = \left(\frac{b+\left(CB_{0f}^2/\mu_0\right)}{b+\left(CB_{0i}^2/\mu_0\right)}\right)^{1/2} \qquad 9.28$$

If the salt is demagnetized to zero applied field, the final temperature reached is

$$T_f = T_i \left(\frac{b}{b+\left(CB_{0i}^2/\mu_0\right)}\right)^{1/2} \qquad 9.29$$

This relation holds for a salt obeying the Curie law and where the heat capacity in a zero magnetic field is given by Equation 9.22. Gadolinium sulfate is one such salt. Some experimental results are presented in Figure 9.3. The linear dependence of $(T_i/T_f)^2$ on B_{0i}^2 is apparent.

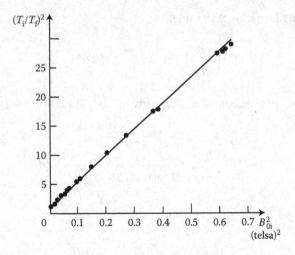

Figure 9.3   Linear dependence of $(T_i/T_f)^2$ with B_{0i}^2 in agreement with Equation 9.29. The measurements are for gadolinium sulfate.

Problems

9.1   Gadolinium sulfate has a magnetic susceptibility in SI units of $+3.5 \times 10^{-3}$. Its molar mass is 603 g/mol and its density is 3.01 g/cm^3. Convert the magnetic susceptibility to mass susceptibility and molar susceptibility.

9.2   Xenon gas has a density of 5.89 kg/m^3 and molar mass 131.3 g. Its molar susceptibility in CGS units is -4.54×10^{-5} cm^3/mol. Convert this value to the normal magnetic susceptibility χ_m in dimensionless SI units.

9.3   It can be shown using statistical analysis (see, e.g., Problem 6.9) that the mean magnetic moment of a paramagnetic sample is $\mu \tanh(\mu B/k_B T)$, where B is the applied magnetic field, T is the temperature, and μ is the magnitude of each individual magnetic moment, which can be either "up" with magnetic moment $+\mu$ or "down" with magnetic moment $-\mu$. Use this result to derive the Curie law.

9.4   In this problem, you will examine the validity of the Curie law at low temperatures. (See the preceding problem.) Assume a paramagnetic material made up of magnetic moments equal to the Bohr magneton: $\mu = \mu_B = e\hbar/2m = 9.27 \times 10^{-24}$ J/T. For an applied field of 3.0 T, find the temperature at which the exact expression $\bar{\mu} = \mu \tanh(\mu B/k_B T)$ and the approximation using $\tanh(\mu B/k_B T) \approx \mu B/k_B T$ differ by 1%. Then, discuss how well the Curie law works at higher temperatures.

9.5   Find the units for magnetic moment $\mathscr{M}$ using Equation 9.5 and show that these units are consistent with those derived from the magnetic moment of a current-carrying loop of wire, equal to the current multiplied by the area of the loop.

9.6   The magnetic susceptibility of aluminum at $T = 273$ K is given by Table 9.1 as 2.1×10^{-5}. (a) At that temperature, what fraction of the maximum magnetic moment is achieved in an applied magnetic field of 5.0 T? Assume that the individual magnetic moments are equal in size to the Bohr magneton, $\mu = \mu_B = e\hbar/2m = 9.27 \times 10^{-24}$ J/T. (b) Repeat using the same applied field if the sample is immersed in liquid nitrogen at $T = 77$ K.

9.7   Use the thermodynamic relations described in Section 9.1.4 to reproduce the three remaining Maxwell relations, in addition to one found in the text.

9.8   A paramagnetic salt is magnetized isothermally and reversibly from zero applied magnetic field to a final value of B_0. It obeys the Curie law $\chi_m = C/T$. Show that the heat of magnetization is

$$Q = -\frac{CV}{T\mu_0}\frac{B_0^2}{2}$$

where V is the volume of the salt. Hint: Express $S = S(T, B_0)$.

9.9   The paramagnetic salt of the previous problem is adiabatically and reversibly demagnetized from an initial state (T_2, B_{01}) to a final state $(T_2, 0)$. The internal energy $U = \alpha T^4$ where α is a constant. Show that

$$T_2^3 = T_1^3 - \frac{3CVB_{01}^2}{8\alpha\mu_0 T_1^2}$$

Hint: A natural way to tackle this problem is to write $T = T(S, B_0)$. Although the problem can be solved in this way, it then involves a very nasty integral. Instead write $T = T(S, \mathscr{M})$ and recognize that $\mathscr{M}$ decreases to zero at the end of the demagnetization.

REFERENCE

Griffiths, D.J., *Introduction to Electrodynamics*, fourth edition, Cambridge University Press, Cambridge, 2017.

Chapter 10: Phase Changes

Everyone is familiar with the fact that, if the temperature is raised, ice melts into water and then the water turns into steam. Ice, water, and steam are the three *phases* of the bulk substance made up of H_2O molecules. It is also a familiar observation that two of the phases can coexist in equilibrium with each other. For example, a beaker at 0°C and at atmospheric pressure can contain ice floating in water with the mass of the ice remaining constant. Similarly, water and steam coexist in a boiling pot at 100°C and at atmospheric pressure. In fact, at one particular temperature and pressure (much lower than 1 atm), all three phases of H_2O may exist together.

Everyday experience gives some sense of the properties that distinguish the three phases of water or any substance. Solids appear rigid outside of any container. Liquids flow freely to the boundaries of a container but have fixed density for a given material and are difficult to compress. Gases also flow—and hence liquids and gases are both called fluids—but unlike liquids, gases tend to expand to fill their container with minimum density. Conversely, gases are also easy to compress, compared with liquids. However, these are all qualitative observations that are sometimes difficult to interpret. For example, what is the phase of warm butter?

Strictly, a phase consists of a homogeneous region of (P, V, T) space, and those regions have definite boundaries; this is certainly so for the ice–water–steam example. The next section will make this distinction more precise.

10.1 *PVT* SURFACES

Consider a system, such as a simple fluid, where P, V, and T are the appropriate state variables. Recall that equilibrium states of the system in a single phase are uniquely specified by any two of these variables, because P, V, and T are connected by the equation of state. Specifying the pair P and V, for example, then fixes T. If these variables are plotted along three mutually perpendicular

DOI: 10.1201/9781003299479-10

axes, the different equilibrium states of the system define a surface, called the *PVT* surface. The *PVT* surface for a typical substance is shown in the center of Figure 10.1.

10.1.1 *PVT* Surfaces and Phases

At first glance, the *PVT* surface appears complicated, principally because it is three-dimensional and thus hard to visualize. It can best be understood as follows. First, consider the isothermal path abcdef in Figure 10.1. Experimentally, one could begin taking the system along this path by compressing it isothermally in a cylinder. At point a, the system is in the single vapor-phase region. As the volume is decreased the pressure increases, until at b some condensation occurs, with drops of liquid just beginning to appear; that is, the substance begins to separate into two distinct phases of quite different densities, although both are at the same temperature and pressure. Moving from b, where the substance is all vapor, to c, more and more liquid appears until, at c, the substance is all liquid; path bc is thus in the two-phase liquid–vapor region. The boundary at b marks a *saturated vapor*, while at c there is a *saturated liquid*.

The path from c to d is in the single-phase liquid region. It requires a great increase in pressure to achieve a small change in volume, because the compressibility of a liquid is generally small.

At d, the substance begins to solidify, until at e it becomes all solid. Path de is in the two-phase solid–liquid region. The path e to f is in the single-phase solid region, where enormous pressures are generally required to effect a compression; thus the slope of ef is very large.

There is a *critical temperature* T_c above which an isothermal compression, such as the one just described, produces no sharp liquid–vapor transition. Upon compression, the system becomes more and more dense, moving continuously from being a low-density fluid into a high-density fluid. The isotherm gh is an example of this. This effect would also be observed if the non-isothermal path, shown as the dotted line starting at a, is followed. Below T_c, as you have seen, it is possible for the system to exist in two separate phases, liquid and vapor, with quite different densities. At the *critical point* C, the vapor and the liquid have become indistinguishable with the same density. It is customary to use the word *gas* above T_c and the word *vapor* below T_c. In other words, compressing a gas will not produce condensation.

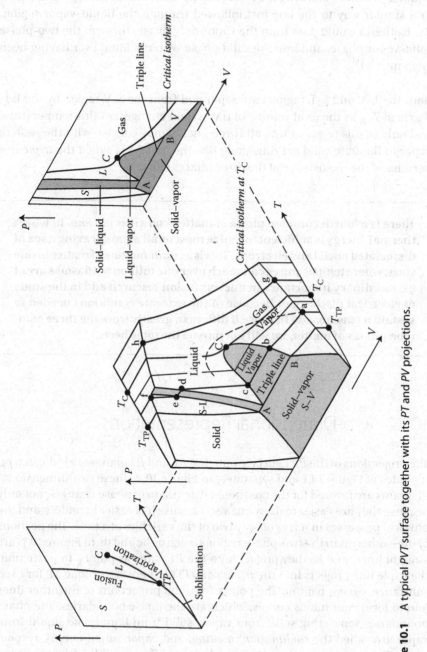

Figure 10.1 A typical PVT surface together with its PT and PV projections.

The region marked S–V represents the two-phase solid–vapor region where no liquid is present. Following the system along an isotherm through this region in a similar way to the one just followed through the liquid–vapor region, the isotherm would pass from the vapor-only phase, through the two-phase solid–vapor phase, and into the solid phase with no liquid ever having been encountered.

Both the L–V and S–L regions are separated from the S–V region by the isotherm at T_{TP}, as the total volume of the system changes. At this temperature, and only at this temperature, all three phases may coexist, with the ratio of vapor to liquid to solid varying along BA. The line BA is called the *triple line* because of the coexistence of the three phases along it.

There is a fourth common phase of matter known as *plasma*, in which thermal energy is sufficient to ionize most or all atoms, leaving a sea of dissociated nuclei and electrons. This is a common state of matter inside stars, where temperatures can reach over one million K. Plasmas are of extraordinary importance in nuclear fusion research and in the study of electrical discharges. Because of the extreme conditions needed to sustain a plasma, and because it differs so greatly from the three common phases of matter, we will not pursue the topic here.

10.1.2 Two-Dimensional Representations

The projections of the PVT surface onto the PV and PT planes are indicated on the sides of Figure 10.1 and separately in Figure 10.2. These two-dimensional diagrams are favored for the continued discussion of phase changes, not only because they are easier to draw but also because it is easier to understand the physical processes in terms of only two of the variables at a time. The portions of the isotherms in the two-phase regions (such as bc and de in Figure 10.1) are straight lines, and so they project onto the PT plane as points. In particular, the triple line projects into the *triple point* (TP); there is a unique TP for each substance, except helium. The points from the projections of the other lines join to form continuous curves, which are the phase boundaries. The phase boundaries separating solid from vapor, solid from liquid, and liquid from vapor are called the *sublimation, melting*, and *vaporization* curves, respectively. These curves are simply the loci of the different sublimation, melting, and boiling points.

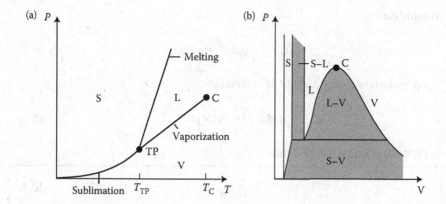

Figure 10.2 (a) The PT projection and (b) the PV projection for various substances. $S =$ solid; $L =$ liquid; and $V =$ vapor.

In the study of thermodynamics, the PT representation is generally favored over the PV representation, and so the PT graph is used throughout this chapter. The two graphs in Figure 10.2 make it clear that the PT view is a simpler, "cleaner" one in that each phase occupies exclusively a simple region of the graph, with mixed phases hidden in the lines. Further, the experimenter is generally able to measure the pressure and temperature accurately. For these reasons, the PT graph is most common.

10.1.3 Equilibrium Condition for Two Phases

Suppose a system consists of two phases of a single substance (also called a single *component*), for example, ice and water, which consist of the single substance H_2O. If this system is maintained at a constant temperature and pressure, you know from Section 7.5 that the condition for thermodynamic equilibrium is that the Gibbs function is a minimum.

In this discussion, we adopt the usual notation that extensive quantities per unit mass, or specific values, take lower case symbols. Thus, g will be used to symbolize the Gibbs function per unit mass. Let the two phases be labeled 1 and 2 with masses M_1 and M_2. Then, the total mass is $M = M_1 + M_2$, and the Gibbs function is

$$G = M_1 g_1 + M_2 g_2 \qquad\qquad 10.1$$

At equilibrium,

$$dG = 0 = g_1 dM_1 + g_2 dM_2 \qquad 10.2$$

If the system is closed so that M is constant,

$$dM = dM_1 + dM_2 = 0 \qquad 10.3$$

Substituting Equation 10.3 into 10.2,

$$g_1 = g_2 \qquad 10.4$$

Therefore, the equilibrium condition for the coexistence of two phases is that the specific Gibbs functions are equal. At equilibrium, any amount M_1 ($<M$) of phase 1 may coexist with the remaining amount $M_2 = M - M_1$ of phase 2 because the value of G is unchanged as M_1 is altered. This explains the existence of the two-phase regions of Figure 10.1. The equality of g for the two phases at equilibrium is a powerful result, which now leads us directly to the Clausius–Clapeyron equation, a general relation of great significance in the study of phase transitions.

10.2 CLAUSIUS–CLAPEYRON EQUATION FOR FIRST-ORDER PHASE CHANGES

A *first-order phase change* in a substance is characterized by a change in the specific volume between the two phases, accompanied by a latent heat. A solid melting into a liquid and a liquid boiling into a vapor are examples of first-order phase changes. First-order phase changes are in fact the familiar type of phase change. These occur along boundary lines in PT projections, as shown in Figure 10.2(a).

10.2.1 Development of the Clausius–Clapeyron Equation

Consider the generic PT projection shown in Figure 10.3. At A, where the pressure and temperature are P and T,

$$g_1(T,P) = g_2(T,P) \qquad 10.5$$

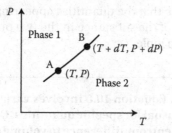

Figure 10.3 A portion of a phase boundary.

At the nearby state B, where the pressure and temperature are $P + dP$ and $T + dT$,

$$g_1(T+dT, P+dP) = g_2(T+dT, P+dP) \qquad 10.6$$

Using Taylor's theorem, Equation 10.6 becomes

$$g_1(T,P) + \left(\frac{\partial g_1}{\partial T}\right)_P dT + \left(\frac{\partial g_1}{\partial P}\right)_T dP = g_2(T,P) + \left(\frac{\partial g_2}{\partial T}\right)_P dT + \left(\frac{\partial g_2}{\partial P}\right)_T dP$$

where it is assumed that for small changes derivatives higher than the first order can be ignored. Using Equation 10.5, this reduces to

$$\left[\left(\frac{\partial g_1}{\partial T}\right)_P - \left(\frac{\partial g_2}{\partial T}\right)_P\right] dT = \left[\left(\frac{\partial g_2}{\partial P}\right)_T - \left(\frac{\partial g_1}{\partial P}\right)_T\right] dP \qquad 10.7$$

From the thermodynamic identity for G,

$$v = \left(\frac{\partial g}{\partial P}\right)_T \quad \text{and} \quad s = -\left(\frac{\partial g}{\partial T}\right)_P \qquad 7.40$$

Substituting these into Equation 10.7,

$$(s_2 - s_1) dT = (v_2 - v_1) dP$$

or

$$\frac{dP}{dT} = \frac{s_2 - s_1}{v_2 - v_1} = \frac{S_2 - S_1}{V_2 - V_1} \qquad 10.8$$

It is important to realize that the quantities appearing in Equation 10.8 must refer to the same mass of the substance in the two phases. For example, this could be a kilogram or a mole.

The first equality in Equation 10.8 involves variables s_i and v_i, which are in lowercase as usual for specific quantities. However, because the equality is a ratio of entropy difference to volume difference, the same ratio holds for any quantity, and hence the last equality with S_i and V_i is valid.

A phase change in a system occurring with a change in its entropy implies that there is a transfer of heat to or from the surroundings; this is the *latent heat L*. If a fixed mass of phase 1 changes into phase 2 at the temperature T, if follows that from Equation 5.4 that $L = T(S_2 - S_1)$. In terms of unit mass,

$$l = T(s_2 - s_1) \qquad \qquad 10.9$$

when $s_2 > s_1$, latent heat l is positive and heat has to be supplied to the system. Substituting Equation 10.9 into 10.8,

---

Clausius – Clapeyron equation:

$$\frac{dP}{dT} = \frac{l}{T(v_2 - v_1)} = \frac{L}{T(V_2 - V_1)} \qquad \qquad 10.10$$

This is the Clausius–Clapeyron equation for the slope of the phase boundary. For this equation to make sense, there has to be a volume change between the two phases as well as a latent heat. These are precisely the conditions for a first-order phase change. The Clausius–Clapeyron equation is a powerful result, because it can be applied to any phase boundary in a first-order phase change, and because it relates the latent heat directly to the state variables P, V, and T.

Figure 10.4(a) shows the PT projection for a substance that expands on melting from the solid to the liquid ($v_L > v_S$) and requiring latent heat for the phase change, so $s_L > s_S$. Then, from Equation 10.10, dP/dT is positive as shown. This is typical of most materials. On the other hand, water contracts when ice melts into liquid water, and so the PT projection is as in Figure 10.4(b) where dP/dT is negative along the solid/liquid boundary. Water belongs to the small class of substances that behave in this way. Note that both for water and for the more

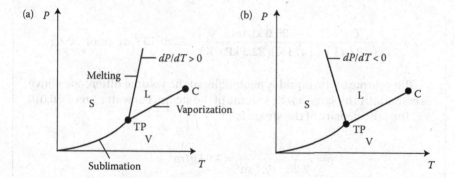

Figure 10.4 (a) A *PT* projection for a typical substance, which expands on melting. (b) A *PT* projection for a substance that contracts on melting, such as water.

typical projection in Figure 10.4(a), the other phase boundaries (solid–vapor and liquid–vapor) have positive slopes throughout, indicating a lower density for the vapor phase. The pressure at any point along one of those other boundaries is called the *vapor pressure*, which clearly tends to be an increasing function of temperature.

When water forms into ice, the molecules join by hydrogen bonding. The nature of these bonds coupled with the unusual 104.5° angle in the H–O–H bond leaves more empty space in ice relative to water, leading to the anomalous density increase.

EXAMPLE 10.1

In a particular electric power plant, water is heated to 200°C to be boiled. At that temperature the latent heat of the phase transition is 35.0 kJ/mol. The vapor pressure of water is 1491 kPa at 198°C and 1621 kPa at 202°C. Use these data to find the volume of one mole of steam and the density of steam at 200°C.

Solution: From the data given, the slope of the phase boundary is

$$\frac{dP}{dT} = \frac{130 \text{ kPa}}{4°C} = 32.5 \text{ kPa/°C}$$

Using the Clausius–Clapeyron equation, the difference in the molar volume between the two phases is

$$\Delta V = \frac{L}{T(dP/dT)} = \frac{35.0 \text{ kJ/mol}}{(473 \text{ K})(32.5 \text{ kPa/K})} = 2.28 \times 10^{-3} \text{ m}^3/\text{mol}$$

The volume of the liquid is negligible, so the volume difference above is essentially the same as the volume of the steam. For water 1 mol = 0.018 kg. Thus the density of the steam is

$$\rho = \frac{0.018 \text{ kg}}{2.28 \times 10^{-3} \text{ m}^3} = 7.9 \text{ kg/m}^3$$

This is very close to the density in published tables.

EXAMPLE 10.2

For the situation described in Example 10.1, what is the entropy change when one mole of water is boiled at 200°C?

Solution: Using the Maxwell relation in Equation 7.26, the change in entropy is

$$\Delta S = \frac{dP}{dT} \Delta V = (32.5 \text{ kPa/K})(2.28 \times 10^{-3} \text{ m}^3) = 74 \text{ J/K}$$

This agrees with published tables.

10.2.2 An Example: Melting Water and Winter Sports

At 0°C, the specific volumes of ice and water are 1.0905×10^{-3} and 1.000×10^{-3} m³/kg, respectively, and the latent heat of fusion is 334 kJ/kg. Substituting these values into Equation 10.10, the slope of the fusion curve is

$$\frac{dP}{dT} = \frac{L}{T(V_2 - V_1)} = \frac{334 \times 10^3 \text{J/kg}}{(273\text{K})(-9.05 \times 10^{-5}\text{m}^3/\text{kg})}$$

$$= -1.35 \times 10^7 \text{Pa/K}$$

or −133 atm/K. It is often claimed that such an extreme negative slope makes it possible to ice skate. This claim should be analyzed carefully.

The bottom of an ice skate is hollow ground, as in Figure 10.5(a), so an enormous pressure is built up under the sharp edge—this is the key to the argument. As a rough estimate, a skate blade that is 30 cm long and 3 mm wide has a cross-sectional area of about 10^{-3} m³, but because of the hollow-ground shape the actual contact area might be say 10% of this, or 10^{-4} m³. Then, the pressure created on this surface by an 800-N skater is 800 N divided by the area (assuming the skater moves on one skate at a time), which is about 8×10^6 Pa or 80 atm.

Now suppose that the temperature is –10°C. At $P = 1$ atmosphere, the ice is in the state α on the PT projection of Figure 10.5(b), and there is no water present. When the ice skater puts pressure on the ice, the state moves along the constant temperature line αγ. In theory, as soon as the phase boundary is reached at β, some ice melts so that the edge of the skate sinks in fractionally, with the load now spread over a larger area, stabilizing the pressure. The state thus remains fixed at β, with the liberated water acting as a lubricant. The question must be asked: Is 80 atm or so sufficient to achieve this goal? From above, $dP/dT = -133$ atm/K, so a pressure increase of $(133$ atm/K$)(10$ K$) = 1330$ atm would be required to go from α to β. Therefore, this explanation for the success of the skater (and the general "slipperiness" of ice) appears inadequate. Although the question is still debated, it is now believed that frictional heating provides a better explanation.

It is often said that skiing and snowboarding also work by the pressure-melting effect. The analysis of this question is left to Problem 10.11. But as a hint, consider that the bottom surfaces of skis or snowboard are made slippery with wax.

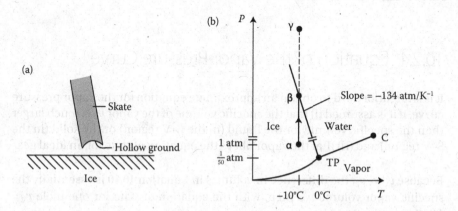

Figure 10.5 The physics of skating: (a) hollow-ground ice skate. (b) PT projection for pure water.

10.2.3 Melting Point of Ice and Boiling Point of Water

It is well known by mountaineers—and even people living in mountainous regions at altitudes above 1 km—that the boiling point of water is affected by the ambient pressure. It is straightforward to analyze this problem using the Clausius–Clapeyron equation.

The specific volumes of water and steam at 100°C and 1 atmosphere are 1.043 × 10⁻³ and 1.673 m³/kg, respectively, and the latent heat of vaporization is 2257 kJ/kg. Substituting into Equation 10.10,

$$\frac{dP}{dT} = \frac{L}{T(V_2 - V_1)} = \frac{2257 \times 10^3 \, \text{J/kg}}{(373\text{K})(1.672\text{m}^3/\text{kg})}$$

$$= +3620 \, \text{Pa/K}$$

or about (1/28) atm/K. On the top of Mount Everest where the pressure is only 0.35 atm, the boiling point of water can be estimated to be depressed by (0.65 atm)/(1/28 atm/K) = 18 K or 18°C, so the boiling point is only about 82°C. In Denver, Colorado the altitude is close to 1.6 km, and typical air pressure is 0.82 atm, and a similar calculation gives a boiling point of 95°C. Even at that altitude, cooking times for foods such as noodles and rice are affected.

> In fact, the *PVT* surface of water is somewhat more complex than has been suggested, but that does not affect the general validity of this discussion.

10.2.4 Equation of the Vapor Pressure Curve

It is straightforward to obtain an approximate equation for the vapor pressure curve, if it is assumed (i) that the specific volume of the vapor v_V is much larger than the specific volume for the liquid (in the L–V region) or the solid (in the S–V region) and (ii) that the vapor obeys the equation of state for an ideal gas.

Because $v_V \gg v_L$, the difference in volumes in Equation 10.10 is essentially the specific vapor volume v_V. Then, with the equation of state for one mole $v_V = RT/P$, Equation 10.10 becomes

$$\frac{dP}{dT} \approx \frac{l}{Tv_V} = \frac{lP}{T^2 R} \qquad\qquad 10.11$$

Integration then gives

$$\ln P = -\frac{l}{RT} + \text{constant} \quad (\text{one mole}) \qquad 10.12$$

where it is assumed that l is constant over the region of the integration.

This equation explains how the vapor pressure varies with temperature. Equation 10.12 predicts the shape of the observed vapor pressure curves in Figure 10.4, which are monotonically increasing and, for all reasonable temperatures $RT < l/2$, also concave upward (see Problem 10.13).

> Equation 10.12 is used in very low-temperature thermometry, at less than 1 K, by relating the measurable pressure above a bath of liquid helium to the temperature. In practice, a pressure gauge is simply calibrated to read the absolute temperature.

10.3 VARIATION OF GIBBS FUNCTION G IN FIRST-ORDER TRANSITIONS

There are some simple analytical arguments from which one can obtain information about the changes of entropy and volume in a first-order phase transition.

First, consider how the specific Gibbs function g varies along the isobar XY in Figure 10.6(a). Along XY, the solid–liquid phase boundary is at temperature T_0. From Section 7.5,

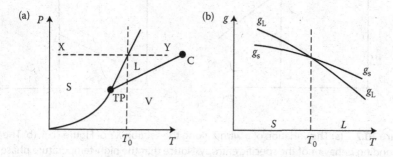

Figure 10.6   (a) An isobaric section XY across a PT projection; (b) It is shown in the text that the Gibbs functions for the two phases vary as shown here.

$$\left(\frac{\partial g}{\partial T}\right)_P = -s \qquad\qquad 7.40$$

Therefore, the g versus T plot at constant pressure has a negative slope because entropy is always positive. Taking the derivative of Equation 7.40,

$$\left(\frac{\partial^2 g}{\partial T^2}\right)_P = -\left(\frac{\partial s}{\partial T}\right)_P = -\frac{c_P}{T} < 0 \qquad\qquad 10.13$$

where the introduction of c_P follows from Equation 7.17, and the inequality is due to the fact that c_P is always positive. Therefore, a plot of g versus T always bends concave downward, toward the T axis. The Gibbs function g is a smoothly varying function of T and P, which can take different values for all T and P. Figure 10.6(b) shows the Gibbs functions for the solid and liquid phases. At T_0, they must be equal, so the curves cross at that point. Also, for $T < T_0$, the solid phase is the stable phase. This means that its g curve must be the lower one in this region, in order to minimize the Gibbs function for the system. The higher g curve for the liquid phase represents the unstable phase. g for the constant P section varies then as in Figure 10.7(a), with a discontinuity in the slope at T_0. This implies a discontinuity in s as shown in Figure 10.7(b), with the higher temperature phase having the greater s.

Another approach is to consider the isotherm X′Y′ of Figure 10.8(a). This time the phase boundary occurs at pressure P_0, as shown. A similar argument gives the plots of g versus P for the solid and liquid phases, as shown in Figure 10.8(b). In this case,

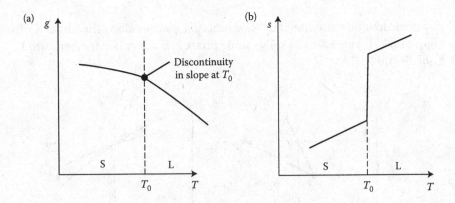

Figure 10.7　(a) The variation of g with T along the section XY of Figure 10.6. (b) The corresponding behavior of the specific entropy. Notice that the high-temperature phase has the higher entropy.

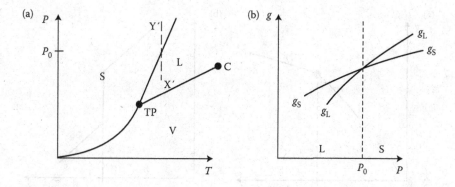

Figure 10.8 (a) An isothermal section X'Y' across a *PT* projection and (b) the text explains why the Gibbs function varies as shown here.

$$\left(\frac{\partial g}{\partial P}\right)_T = v > 0 \qquad\qquad 7.40$$

because volume is positive. Taking another derivative,

$$\left(\frac{\partial^2 g}{\partial P^2}\right)_T = \left(\frac{\partial v}{\partial P}\right)_T < 0$$

where the inequality follows from an empirical fact: all known substances suffer a decrease in volume upon an increase of pressure. Thus, the Gibbs function varies as in Figure 10.9(a), with a discontinuity in the slope at $P = P_0$. From this, it follows that there is a corresponding discontinuity in volume. This implies that v varies as in Figure 10.9(b), with the high-pressure phase having the smaller specific volume. This result is consistent with experience.

10.4 SECOND-ORDER PHASE CHANGES

For a first-order phase change, the following facts hold true. There is a change in the specific volume; there is a latent heat, which means that there is a change in the specific entropy; and there is no change in the specific Gibbs function. That is,

$$g_1 = g_2, v_1 \neq v_2, l \neq 0, \text{ and } s_1 \neq s_2$$

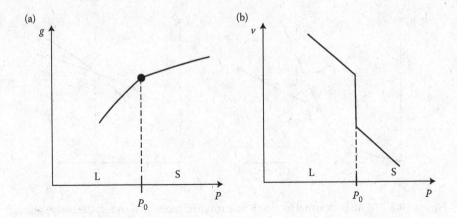

Figure 10.9 (a) The variation of g with P along the section X'Y' of Figure 10.8. (b) The corresponding behavior of the specific volume. Notice that the high-pressure phase has the smaller specific volume, as expected.

Recalling that

$$s = -\left(\frac{\partial g}{\partial T}\right)_P \quad \text{and} \quad v = \left(\frac{\partial g}{\partial P}\right)_T \qquad\qquad 7.40$$

a first-order phase change is one in which g is continuous, but the first-order derivatives of g with respect to the natural variables P and T are discontinuous.

Second-order phase changes can be defined by extending this classification. In a second-order phase change, g is continuous; the first-order derivatives of g with respect to T and P are now continuous; therefore, there is no change in the specific volume and there is no latent heat. However, there are discontinuities in the second-order derivatives. That is, g, s, and v are continuous, but

$$\left(\frac{\partial s}{\partial T}\right)_P, \left(\frac{\partial s}{\partial P}\right)_T, \left(\frac{\partial v}{\partial T}\right)_P, \text{ and } \left(\frac{\partial v}{\partial P}\right)_T$$

are discontinuous. Recall that the heat capacity, coefficient of thermal expansion, and compressibility are given by

$$c_P = T\left(\frac{\partial s}{\partial T}\right)_P, \quad \beta = \frac{1}{v}\left(\frac{\partial v}{\partial T}\right)_P, \text{ and } \kappa = -\frac{1}{v}\left(\frac{\partial v}{\partial P}\right)_T,$$

This means that c_P, β, and κ are discontinuous in second-order phase changes. The second and third partial derivatives are the same, by the Maxwell relation Equation 7.41, apart from a difference in sign, which is of no importance to the question of continuity.

This classification of phase changes can be further extended to third and even higher order phase changes and is known as the *Ehrenfest classification*.

10.5 EXAMPLES OF PHASE CHANGES OF DIFFERENT ORDERS

There are many examples of first- and second-order phase changes, or transitions, in physics, metallurgy, and chemistry. Below are examples of a few in each category.

10.5.1 First-Order Phase Changes

1. A solid melting into a liquid, a liquid boiling into a gas, and a solid subliming into a gas are the most familiar examples of a first-order phase change.

2. Below a certain characteristic temperature, the critical temperature T_c, some materials become superconductors. They behave as normal conductors at temperatures above T_c, exhibiting electrical resistance, but below T_c they have zero electrical resistance, as shown in Figure 10.10. A current circulating in a ring of pure lead would go on circulating forever, provided the lead was kept cold at a temperature less than $T_c = 7.2$ K. The values for T_c for a few superconductors are given in Table 10.1.

The superconducting phase can be destroyed by raising the temperature above T_c in the absence of a magnetic field. At any $T < T_c$, the superconducting phase may also be destroyed by an applied magnetic field above a certain critical value of $B_c(T)$; this critical field increases as the temperature is reduced, reaching the value $B_c(0)$ as $T \rightarrow 0$. This is illustrated for a typical superconductor in the phase diagram shown in Figure 10.11, where the phase boundary

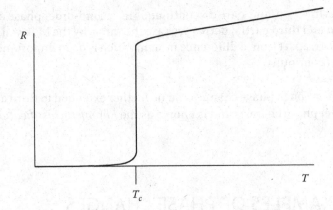

Figure 10.10 Resistance as a function of temperature for a superconductor. Below temperature T_c, the resistance is zero.

TABLE 10.1   Values of Critical (Transition) Temperature T_c for a Few Superconductors

Superconductor	Critical Temperature T_c (K)	Critical Field $B_c(0)$ (T)
Tin	3.7	0.031
Mercury	4.2	0.042
Lead	7.2	0.081
Niobium	9.2	0.26
Nb_3Sn	18.1	29[a]
$YBa_2Cu_3O_7$	93	250[a]
$HgBa_2Ca_2Cu_4O_{1+x}$	134	>100[a]

[a] Type II superconductors, with an upper critical field at which electrical resistance vanishes and a lower critical field at which magnetic flux is expelled. Values given are the upper critical field.

separates the normal from the superconducting phase. The third column of Table 10.1 gives the values of $B_c(0)$ for different superconductors.

The transition from a superconductor to a normal conductor is a first-order change, provided the transition takes place in an applied magnetic field. The arrow 1 indicates such a change in Figure 10.11.

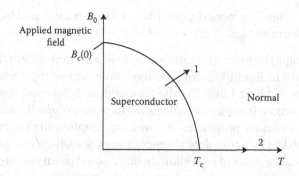

Figure 10.11 A phase diagram for a typical superconductor. The transition 1, occurring in an applied magnetic field, is a first-order phase transition. The transition 2, occurring due to changing temperature in the absence of a magnetic field, is an example of a second-order phase transition.

Niobium has the highest transition temperature of any elemental superconductor. In the 1980s, superconductors such as $YBa_2Cu_3O_7$ were discovered with transition temperatures above 77 K, the boiling point of nitrogen at 1 atm. This raised the possibility of more practical applications and generated a search for materials with even higher transition temperatures. A room-temperature superconductor has not been found, but if it were there would be enormous energy savings in electrical transmission and other applications.

10.5.2 Second-Order Phase Changes

1.   Magnetic materials fall broadly into three main classes; the *diamagnetic class*, where the atoms of the material have no permanent magnetic moment but one is induced on the application of a magnetic field; the *paramagnetic class*, where the atoms have a permanent magnetic moment but it requires the application of an external magnetic field to overcome the random orientation of the individual moments and to give the material a net magnetic moment; and the *ferromagnetic* class, where the atoms have a net magnetic moment, which couple together to give the material a net moment even in the absence of an external magnetic field. As the temperature of a ferromagnet is raised, it becomes a paramagnet at the *Curie temperature*. That transition is a second-order phase change.

2.   The transition of a superconductor to a normal conductor is a second-order phase change, provided the transition does not take place in an applied magnetic field. Then, as described above, the transition is of

first order. The second-order phase change is indicated by the arrow 2 in Figure 10.11.

3.   Perhaps the most dramatic example of a second-order phase change occurs in liquid helium. The most common isotope ⁴He is a liquid below 4.2 K at 1 atm. As the temperature is lowered through 2.2 K, it changes from a normal liquid to a *superfluid* liquid with quite extraordinary properties. As the name implies, the superfluid phase is marked by a complete absence of any viscosity, or resistance to flow. A rotating mass of superfluid helium would go on rotating forever, in exact analogy to the persistence of the current in a ring of superconductor, provided that the velocity is below a certain critical limit.

To describe superfluid helium in more detail, some new notation is needed. Helium can exist as two isotopes: the common isotope is ⁴He, commonly called helium four; and the rare lighter isotope is ³He, commonly called helium three. Only the common isotope ⁴He undergoes the superfluid transition at 2.2 K. Unfortunately, the two phases are confusingly called Helium I for the normal phase and Helium II for the superfluid phase, a numbering system that has nothing to do with the mass number.

A plot of the heat capacity of ⁴He against temperature is shown in Figure 10.12. The discontinuity in the heat capacity at 2.2 K is clear. Because of the resemblance of this curve to the Greek letter lambda, the transition temperature is called the *lambda point*.

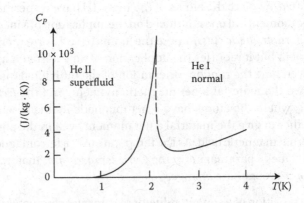

Figure 10.12   The λ-point anomaly in the heat capacity of ⁴He.

> **Helium 3 is a rare isotope, found at barely more than one part per million in naturally occurring helium. It also exhibits superfluid behavior, but only at extremely low temperatures below about 3 mK.**

10.5.3 Ehrenfest Equations for Second-Order Phase Changes

There exist two simple relations for the slope dP/dT of the phase boundary in a second-order phase change. It is possible to derive these two relations in an analogous manner to that used in deriving the Clausius–Clapeyron equation for the slope of the phase boundary in a first-order phase change. The Clausius–Clapeyron equation does not give the slope of the phase boundary for a second-order transition, because both l and Δv vanish, with Equation 10.10 giving an indeterminate answer for dP/dT.

Consider Figure 10.3 again, where there is now a second-order phase change between the two phases, and the solid line is the phase boundary for this transition. As before, consider the two neighboring points A and B on the phase boundary at (T, P) and $(T + dT, P + dP)$. To begin, recall that in a second-order phase change there is no change in either s or v in going from one phase to another. Therefore,

$$\text{At A, } s_1(T,P) = s_2(T,P) \tag{10.14}$$

$$\text{At B, } s_1(T+dT,P+dP) = s_2(T+dT,P+dP) \tag{10.15}$$

Using Taylor's theorem and keeping only first-order derivatives for small changes, Equation 10.15 becomes

$$s_1(T,P) + \left(\frac{\partial s_1}{\partial T}\right)_P dT + \left(\frac{\partial s_1}{\partial P}\right)_T dP = s_2(T,P) + \left(\frac{\partial s_2}{\partial T}\right)_P dT + \left(\frac{\partial s_2}{\partial P}\right)_T dP$$

This can be simplified using Equation 10.14

$$\left(\frac{\partial s_1}{\partial T}\right)_P dT + \left(\frac{\partial s_1}{\partial P}\right)_T dP = \left(\frac{\partial s_2}{\partial T}\right)_P dT + \left(\frac{\partial s_2}{\partial P}\right)_T dP \tag{10.16}$$

Multiplying Equation 10.16 through by T and remembering that

$$c_P = T\left(\frac{\partial s}{\partial T}\right)_P, \quad \beta = \frac{1}{V}\left(\frac{\partial V}{\partial T}\right)_P = \frac{1}{v}\left(\frac{\partial v}{\partial T}\right)_P, \quad \text{and } v_1 = v_2$$

and the Maxwell relation

$$\left(\frac{\partial s}{\partial P}\right)_T = -\left(\frac{\partial v}{\partial T}\right)_P \qquad 7.41$$

Equation 10.16 reduces to

$$c_{P1}dT - Tv\beta_1 dP = c_{P2}dT - Tv\beta_2 dP \qquad 10.17$$

where the subscripts on c_p and β refer to the two phases. Solving for dP/dT in Equation 10.17,

$$\frac{dP}{dT} = \frac{c_{P1} - c_{P2}}{Tv(\beta_1 - \beta_2)} = \frac{C_{P1} - C_{P2}}{TV(\beta_1 - \beta_2)} \qquad 10.18$$

This is the *first Ehrenfest equation*. As in the case of the Clausius–Clapeyron equation, the extensive quantities (here C_P and V) must refer to the same mass of the substance in each phase.

Now consider the continuity of v in a second-order transition

$$\text{At A, } v_1(T,P) = v_2(T,P) \qquad 10.19$$

$$\text{At B, } v_1(T+dT, P+dP) = v_2(T+dT, P+dP) \qquad 10.20$$

Following the same procedure as above, applying Taylor's theorem to Equation 10.20 and keeping only first-order derivatives gives

$$v_1(T,P) + \left(\frac{\partial v_1}{\partial T}\right)_P dT + \left(\frac{\partial v_1}{\partial P}\right)_T dP = v_2(T,P) + \left(\frac{\partial v_2}{\partial T}\right)_P dT + \left(\frac{\partial v_2}{\partial P}\right)_T dP$$

Using Equation 10.19, this simplifies to

$$\left(\frac{\partial v_1}{\partial T}\right)_P dT + \left(\frac{\partial v_1}{\partial P}\right)_T dP = \left(\frac{\partial v_2}{\partial T}\right)_P dT + \left(\frac{\partial v_2}{\partial P}\right)_T dP \qquad 10.21$$

Remembering that

$$\beta = \frac{1}{V}\left(\frac{\partial V}{\partial T}\right)_P = \frac{1}{v}\left(\frac{\partial v}{\partial T}\right)_P \quad \text{and} \quad \kappa = -\frac{1}{V}\left(\frac{\partial V}{\partial T}\right)_T = -\frac{1}{v}\left(\frac{\partial v}{\partial P}\right)_T$$

Equation 10.21 becomes

$$\beta_1 v_1 dT - \kappa_1 v_1 dP = \beta_2 v_2 dT - \kappa_2 v_2 dP \qquad 10.22$$

Collecting terms, and remembering that $v_1 = v_2$,

$$\frac{dP}{dT} = \frac{\beta_2 - \beta_1}{\kappa_2 - \kappa_1} \qquad 10.23$$

This is the *second Ehrenfest equation*.

A good example of the application of the first Ehrenfest equation is in determining the slope of the phase boundary between He I and He II. Figure 10.13 is the phase diagram for ^{4}He, where the λ line is the phase boundary for the second-order phase change between the two liquid phases (normal and superfluid). Measurements give the values of c_P and β for the two phases. Equation 10.18 then gives results in good agreement with value of the slope determined in other ways (see Problem 10.15).

> **Notice on Figure 10.13 that the solid phase is only reached upon the application of pressures of 25 atm or greater.**

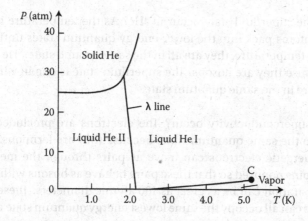

Figure 10.13 Phase diagram for ^{4}He. The normal liquid phase is called He I, and the superfluid phase is called He II.

10.6 SUPERCONDUCTIVITY AND SUPERFLUIDITY

This chapter concludes with a brief account of the theoretical explanation of superconductors and superfluids, including some parallels between them. You have already seen that the persistent flow in a rotating superfluid resembles the persistent current flow in a superconducting ring.

According to quantum mechanics, particles in a restricted geometry exist in discrete quantum states with different energies, as in Figures 10.6–10.9. In the condensed state, under conditions of high particle number density and low temperatures, quantum effects become manifestly important, because at high particle densities there is significant overlap in the quantum-mechanical wave functions of neighboring particles. This is particularly so for electrons in a metal and the light atoms of liquid ^{4}He and liquid ^{3}He.

Because of the indistinguishability of atomic and subatomic particles, such particles can be either *bosons or fermions*. There is no restriction on the number of bosons that can occupy a single quantum state; however, only one fermion can occupy each quantum state. Atomic particles have the property of *spin*, which is a measure of their intrinsic angular momentum. Particles with half-integer spin behave as fermions, and those with integer spin (including zero) behave as bosons. Electrons, which have half-integer spin, are thus fermions. In contrast, the nuclei of ^{4}He, which have zero spin (two protons with opposed spins and two neutrons also with opposed spins), are bosons. However, the nuclei of ^{3}He, with one less neutron than ^{4}He, have half-integer spin and therefore form a fermion system.

How does the superfluid state occur in ^{4}He? As the temperature is lowered, all the ^{4}He atoms pack into the lower-energy quantum states until, at a sufficiently low temperature, they are all in the same ground state. ^{4}He atoms can do this because they are bosons. The superfluid state is one in which all the ^{4}He atoms are in the same quantum state.

How does superconductivity occur? The electrons are precluded from all packing into the same quantum state, because they are fermions. However, the most energetic electrons can move as pairs through the metal lattice, with their spins opposed so that these pairs behave as bosons with zero spin. The pair separation can be as great as 100 atomic diameters. These so-called *Cooper pairs* can all occupy the same lowest energy quantum state at low temperatures, and this gives rise to superconductivity.

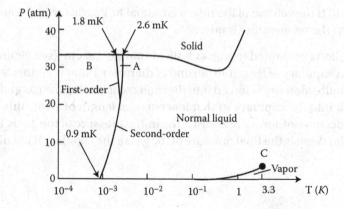

Figure 10.14 Phase diagram for ³He.

The question now arises as to whether a similar pairing of the nuclei of ³He can occur, so that these ³He pairs behave as a boson system, giving rise to superfluid phase. The theory here preceded the experimental evidence. Theory shows that such pairings can occur, except that the half-integer spins add to give a spin of one for the pair and thus a boson system. The superfluid phase in ³He was found at 3 a.m. on Thanksgiving Day at Cornell University in 1971, but not until the temperature had been reduced to 2.6 mK. In fact, a second superfluid phase was also found at the even lower temperature of 1.8 mK. The Cornell experimenters were actually following the solid–liquid transition as a function of temperature when they made their discovery.

The phase diagram for ³He is shown in Figure 10.14. The boundary line between the normal liquid and the two superfluid phases A and B denotes a second-order transition, while the boundary between the two superfluid phases denotes a first-order transition. Below 1 mK, the vapor pressure is so small that it cannot be shown on this scale.

Further discussion of fermions and bosons is given in Chapter 13.

Problems

10.1 A small amount of pure liquid is contained in a glass tube from which all the air has been removed (Figure 10.15(a)). The volume of the tube is significantly greater than the critical volume V_c of the enclosed liquid. (a) Describe what happens as the temperature is raised so that the substance traverses the path XY on the PV projection in Figure 10.15(b).

(b) If the volume of the tube were equal to V_c, what would you observe as the temperature is raised?

10.2 Gas is contained in a glass bulb of volume 250 cm³ (see Figure 10.16). A capillary of length 10 cm and of diameter 1 mm is connected to the bulb. Mercury is forced into the bulb compressing the gas and forcing it into the capillary so that it occupies a length of 1 cm. This process occurs isothermally at 20°C. The initial pressure of the gas is 10^{-3} torr. (a) What is the final pressure of the gas in the capillary if it is nitrogen?

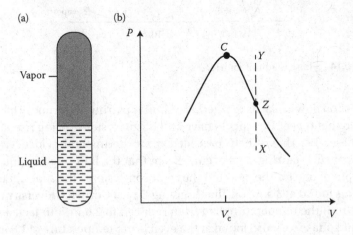

(a)   (b)

Vapor

Liquid

Figure 10.15

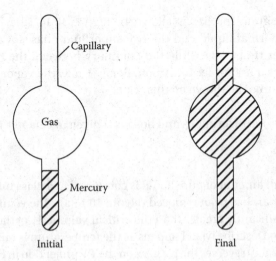

Capillary

Gas

Mercury

Initial   Final

Figure 10.16

(b) What is the final pressure of the gas in the capillary if it is water vapor? (c) How much water condenses? Justify any assumptions. (The vapor pressure of water at 20°C is 17.5 torr.)

10.3 Consider the isotherm at T on the PV projection shown in Figure 10.17. At the point K, the substance is a mixture of liquid and vapor. Let the masses of liquid and vapor be m_1 and m_v and the total mass of the substance be m. Then, the volume occupied by the mixture at K is $m_1 v_1 + m_v v_v$, where v_1 and v_v are the specific volumes of the liquid and vapor. Let the specific volume of the mixture at K be v. Show that

$$m_1(v - v_1) = m_v(v_v - v)$$

This result, which gives the ratio m_v/m_1, is known as the "lever rule," for obvious reasons. The ratio m_v/m_1 is called the *quality* of the mixture.

10.4 At the critical point, $(\partial P/\partial V)_T = 0$ and $(\partial^2 P/\partial V^2)_T = 0$. Show that, for a van der Waals gas (see Section 3.5.4), the critical point is at

$$P_c = \frac{a}{27b^2}; \quad V_c = 3nb; \quad T_c = \frac{8a}{27Rb}$$

10.5 The vapor pressure of camphor is as follows:

Temperature (°C)	30.8	55.0	62.0	78.0
Pressure (Torr)	1.04	3.12	4.22	7.8

By plotting $\ln P$ as a function of $1/T$, estimate the latent heat of vaporization.

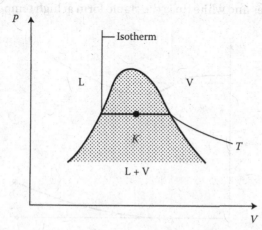

Figure 10.17

10.6 The equations of the sublimation and the vaporization curves of a particular material are given by

$$\ln P = 0.04 - \frac{6}{T} \text{ for sublimation}$$

$$\ln P = 0.03 - \frac{4}{T} \text{ for vaporization}$$

where P is in atmospheres. (a) Find the temperature and pressure of the triple point. (b) Show that the molar latent heats of vaporization and sublimation are $4R$ and $6R$. You may assume that the specific volume in the vapor phase is much larger than those in the liquid and solid phases. (c) Find the latent heat of fusion. Hint: Consider a loop round the triple point in the PT projection. Because S is a state function,

$$\left[\frac{l_{SV}}{T_{TP}}\right] - \left[\frac{l_{SL}}{T_{TP}}\right] - \left[\frac{l_{LV}}{T_{TP}}\right] = 0.$$

10.7 The phase diagram for ^{3}He is as in Figure 10.18. Discuss the variation of G along the isobar XY. What does this tell you about the entropy of the solid phase compared with the entropy of the liquid phase at the lowest temperature? The result should surprise you.

10.8 Tin can exist in two forms, gray tin (also called alpha, a brittle and non-conductive form) and white tin (also called beta, the metallic form that conducts electricity). Gray tin is the stable form at low temperatures and white tin is the stable form at high temperatures. There

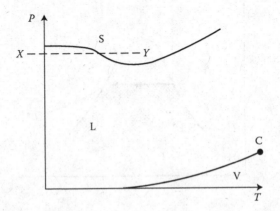

Figure 10.18

is a first-order transition between the two phases with a transition temperature of 286 K at a pressure of 1.00 atm. What is the change in this transition temperature if the pressure is increased to 100 atm? (The latent heat for the transition is 2.20×10^3 J/mol. The densities of gray and white tin are 5.77×10^3 kg/m^3 and 7.37×10^3 kg/m^3. The atomic weight of tin is 118.7 g/mol.)

10.9 The Curie temperature for nickel for the phase change from the ferromagnetic phase to the paramagnetic phase is 630 K at a pressure of 1.0 atm. If the pressure is increased by 100 atm, calculate the shift in the Curie temperature. [In this phase transition, c_P changes by 6.7 J/(K·mol) and β changes by 5.5×10^{-6} K^{-1}. The density of nickel is 8.91×10^3 kg/m^3, and its atomic weight is 58.7 g/mol.]

10.10 As an alternative derivation of the Clausius–Clapeyron equation, begin by using the fact that $g_1 = g_2$ along the phase boundary shown in Figure 10.3. Then apply the thermodynamic identity for G (Equation 7.38) and solve for dP/dT.

10.11 Consider sliding along the icy surface of snow on a mountain using skis or a snowboard. Make reasonable estimates of the surface area of the skis or snowboard and the mass of the person on board. Use those estimates to address the question of whether the pressure on the surface forms a thin layer of water that helps the person glide with reduced friction.

10.12 The two common phases of pure carbon are diamond and graphite. At 1.0 atm and 298 K, graphite is more stable, because its Gibbs function is lower by about 2.9 kJ/mol. (a) Using the densities (3520 kg/m^3 for diamond and 2260 kg/m^3 for graphite), find the pressure required for diamond to become more stable. (b) If this process occurs naturally under rock with density of 3000 kg/m^3, what depth of rock is required? (Assume no change in temperature.)

10.13 Use Equation 10.12 to argue that the vapor pressure curve should be concave upward, as it is in Figure 10.4 for both sublimation and boiling.

10.14 Using Equation 10.12, graph the vapor pressure curve for water from 50°C to 150°C. Assume that the latent heat is constant at the value it assumes at 100°C, which is 2260 kJ/kg. Compare the shape of the curve with the one shown in Figure 10.4(b).

10.15 Use the first Ehrenfest equation to compute the slope of the transition line dP/dT for liquid helium at 1.0 atm and 2.17 K. Use the following data: density = 147 kg/m^3; expansion coefficient = +0.022 K^{-1} (normal) and −0.043 K^{-1} (superfluid); specific heat = 8.5 J/(g·K) (normal) and 16.0 J/(g·K) (superfluid). Compare with the measured value for the slope −78 atm/K.

10.16 The vapor pressure of water is 94.4 kPa at 98°C and 108.9 kPa at 102°C. The density of liquid water at 100°C is 958 kg/m^3. (a) What is the density of steam at that temperature? (b) What is the entropy change for 1 kg of water when it boils at 100°C?

10.17 For boiling water at 1.0 atm and 100°C, the latent heat is 40.7 kJ/mol. Use these data along with Equation 10.12 to graph the vapor pressure curve from 90°C to 100°C, assuming a constant latent heat. Find the boiling point of water at Denver, where $P = 0.82$ atm.

10.18 Carbon dioxide has no liquid phase at atmospheric pressure. Instead, it sublimates from a solid to vapor at −78.6°C, with latent heat 573 kJ/kg. The vapor pressure at a slightly lower temperature, −80.0°C, is 0.89 atm. What is the density of CO_2 gas when it sublimates at −78.6°C? Compare your result with the ideal gas law.

BIBLIOGRAPHY

Annett, J., *Superconductivity, Superfluids, and Condensates*, Oxford University Press, Oxford, 2004.

Satterly, J., The physical properties of solid and liquid helium. *Reviews of Modern Physics 8*, 347–357, 1936.

Tinkham, M., *Introduction to Superconductivity*, second edition, Dover, Mineola, NY, 2004.

Chapter 11: Open Systems and Chemical Potential

There are many physical systems in which the quantity of matter is not fixed. These are called *open systems*. This short chapter will show how such systems are treated in thermodynamics by giving a brief introduction to the concept of chemical potential.

11.1 CHEMICAL POTENTIAL

As an example of such a system, imagine a block of ice floating in water (Figure 11.1). As the ice melts, its mass decreases because there is a transfer of H_2O molecules across the phase boundary dividing the ice from the liquid water. Another example is a chamber containing a small hole through which gas may enter from or leave to the surroundings; the gas in the chamber is then a system of variable mass.

11.1.1 Chemical Potential and Internal Energy U

For a variable-mass system such as the gas-filled chamber with a hole in a wall, it is necessary to modify the thermodynamic identity

$$dU = TdS - PdV \qquad\qquad 5.10$$

to allow for the energy change of the system due to the addition or removal of particles. This extra energy is clearly of importance if it can be released to the rest of the system, and this could be the case if the particles were involved in, say, a chemical reaction of some form. Suppose for a moment that the system consists of only one type of particle, and dN particles are added. Then Equation 5.10 can be modified as

DOI: 10.1201/9781003299479-11

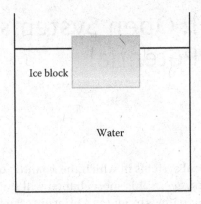

Figure 11.1   Block of ice floating in water is a common example of an open system.

$$dU = TdS - PdV + \mu dN \qquad\qquad 11.1$$

where μ is called the *chemical potential*, defined as the increase in the internal energy per particle added under conditions of constant S and V. That is,

$$\mu = \left(\frac{\partial U}{\partial N}\right)_{S,V} \qquad\qquad 11.2$$

If there is more than one type of particle, Equation 11.1 has to be modified (see Problem 11.1) to

$$dU = TdS - PdV + \sum_i \mu_i dN_i \qquad\qquad 11.3$$

where the chemical potential for the ith type of particle is $\mu_i = (\partial U/\partial N_i)_{S,V,N_k}$ and the symbol N_k means that all the other Ns except N_i are held constant. It is best to assume for the moment that there is only one type of particle present and extend the argument to the more general case of the presence of different types of particles only when this is appropriate.

Even if a particular type of particle is initially absent from the system, that does not mean that the corresponding μ is zero, because it is a measure of the effect on U brought about by the addition of that type of particle while forcing V and S of the system to remain constant. It is the forcing of these two quantities to be constant despite a particle being added that results in μ being less than, equal to, or greater than zero, depending on the system and state under consideration.

Equation 11.3 is an important result. It is a more general form of the thermodynamic identity presented in Equation 7.2. Equation 11.3 is more general because it allows for changes in internal energy, volume, and particle number. This same approach may be applied to the other thermodynamic identities from Chapter 7, involving *H*, *F*, and *G*.

11.1.2 Chemical Potential and Helmholtz Function *F*

Although the definition of μ has a clear physical interpretation, Equation 11.2 is often inconvenient because of the requirement of constant entropy. Another definition of μ can be given in terms of the Helmholtz function *F*. *F* is related to *U* and *S* by

$$F = U - TS \qquad\qquad 7.21$$

After differentiating as in Section 7.4:

$$dF = dU - TdS - SdT \qquad\qquad 7.22$$

it follows using Equation 11.1 that

$$dF = -PdV - SdT + \mu dN \qquad\qquad 11.4$$

Therefore

$$\mu = \left(\frac{\partial F}{\partial N}\right)_{V,T} \qquad\qquad 11.5$$

That is, μ is the increase of the Helmholtz free energy upon the addition of one particle under conditions of constant *T* and *V*.

The connection between μ and *F* may be expanded using the relationship between *F* and the statistical partition function *Z*:

$$F = -Nk_B T \ln Z \qquad\qquad 7.36$$

Using this form of F in the expression for μ in Equation 11.5 gives

$$\mu = -k_B T \left(\frac{\partial}{\partial N}(N \ln Z) \right)_{V,T} \qquad 11.6$$

11.1.3 Chemical Potential and Gibbs Function G

By a similar process, chemical potential can be linked to the Gibbs function G. Recall from Equation 7.37 that

$$G = H - TS = U + PV - TS$$

where the second equality uses the fact that $H = U + PV$ (Equation 3.7). Differentiating,

$$dG = dU + PdV + VdP - TdS - SdT$$

Using Equation 11.1, this simplifies to

$$dG = VdP - SdT + \mu dN \qquad 11.7$$

Therefore

$$\mu = \left(\frac{\partial G}{\partial N} \right)_{T,P} \qquad 11.8$$

This is another useful way to think of the chemical potential. In order words, μ is the increase of the Gibbs free energy upon the addition of one particle under conditions of constant T and P.

Equations 11.2, 11.5, and 11.8 give three independent ways to define chemical potential. Notice that in each case the definition of μ is the partial derivative of a thermodynamic potential (U, F, and G, respectively) with respect to the particle number, with the appropriate natural variables held constant.

The definition of μ in terms of G (Equation 11.8) is of particular importance. Note that G is an extensive quantity, which means that it must be proportional to the particle number. Therefore, it is possible to express G as

$$G(T,P,N) = N\phi(T,P) \qquad 11.9$$

where $\phi(T, P)$ depends on the particular system being considered. Notice that $\phi(T, P)$ does not depend on N, because it is a function of only *intensive* parameters. Differentiating G with respect to N, keeping P and T constant,

$$\mu = \phi(T,P) \qquad \qquad 11.10$$

which is independent of N. Also, it follows from Equation 11.9 that

$$\mu = \left(\frac{G}{N} \right) \text{ (one type of particle present)} \qquad 11.11$$

Thus, μ is simply the Gibbs free energy per particle, provided only one type of particle is present.

This is a more useful statement than the one given in Equation 11.8, which refers to the incremental increase in the Gibbs free energy per particle under conditions of constant T and P. In general, when there are a number of different types of particle present (for example in air with N_2, O_2, and small amounts of other gases), this incremental increase depends on the existing particle populations. In that case Equation 11.11 must be modified, as shown below in Equation 11.16.

On the other hand, it is impossible to obtain similar simple results for μ in terms of U and F. μ is neither U/N nor F/N, because the natural variables for U and F are not both intensive as they are for G. For example, consider U. Because U is extensive, the analog to Equation 11.9 is

$$U = U(S,V,N) = N\phi'\left(\frac{S}{N}, \frac{V}{N} \right) \qquad 11.12$$

where the new function ϕ' depends on the entropy per particle S/N and volume per particle V/N. By Equation 11.2, differentiating with respect to N at constant S and V gives μ:

$$\mu = \phi' + N\left(\frac{\partial \phi'}{\partial N} \right)_{S,V} = \frac{U}{N} + N\left(\frac{\partial \phi'}{\partial N} \right)_{S,V} \qquad 11.13$$

Therefore, μ is not simply U/N but instead involves extra terms that depend on the particle number.

As another way to see the special relationship between μ and G, note that the quantities U, S, V, and N_i in Equation 11.3 are all extensive. Therefore, if the system is scaled up in size by a factor $(1 + \alpha)$, then each of the extensive quantities increases by the same proportion. Suppose now that $\alpha \ll 1$, so that $\alpha = \Delta U/U \approx dU/U$. Similarly for the other extensive quantities in the limit of small α:

$$\alpha = \frac{dU}{U} = \frac{dS}{S} = \frac{dV}{V} = \frac{dN_i}{N_i} \qquad 11.14$$

for all i. Applying these results to each term in Equation 11.3,

$$U = TS - PV + \sum_i \mu_i N_i \qquad 11.15$$

where the α in each term has been canceled.

Equation 11.15 can be related to the Gibbs function by recalling that

$$G = H - TS \qquad 7.37$$

and

$$H = U + PV \qquad 3.7$$

Therefore $G = U + PV - TS$, and by comparison with Equation 11.15

$$G = \sum_i \mu_i N_i \qquad 11.16$$

Equation 11.16 is the generalization of an earlier result, Equation 11.11. It reduces to Equation 11.11 if all the N_i except one are set equal to zero.

11.1.4 Chemical Potential and Equilibrium

The concept of chemical potential is a particularly useful tool for considering equilibrium between two systems that are allowed to exchange particles. For simplicity, the discussion in this section will focus on two systems consisting of different phases of the same substance. Thus, there will be a single particle type throughout both systems. A good example of this is ice melting in water (Figure 11.1), in which there is an exchange of particles (water molecules) between the two systems. What are the general conditions for the two phases to be in equilibrium against this particle exchange?

Figure 11.2 shows two phases of the same substance, labeled A and B, occupying a chamber with rigid adiabatic walls. The two phases occupy volumes V_A and V_B and are separated by a phase boundary. Keeping in mind the example of ice and water, there may be heat and particle flow across the boundary, and the boundary moves as ice melts or water freezes. Phase A has N_A particles, volume V_A, internal energy U_A and pressure P_A, with similar quantities N_B, V_B, U_B, and P_B defined for phase B. These quantities are subject to conservation conditions:

$$N_A + N_B = N \text{ (total number of particles)} \qquad 11.17$$

$$V_A + V_B = V \text{ (total volume)} \qquad 11.18$$

$$U_A + U_B = U \text{ (total internal energy)} \qquad 11.19$$

By the first law, U is fixed because the chamber has adiabatic walls. At equilibrium, the entropy for the thermally isolated combined system of the two phases is a maximum:

$$S = S(U_A, V_A, N_A, U_B, V_B, N_B)$$

$$= S_A(U_A, V_A, N_A) + S_B(U_B, V_B, N_B)$$

$$= \text{a maximum}$$

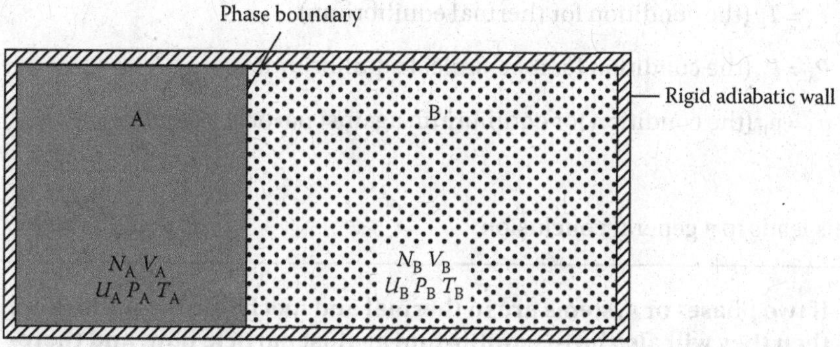

Figure 11.2 Chamber containing two phases of the same substance. There can be particle exchange and heat flow across the boundary. The diathermal boundary can move freely, so the two volumes are not fixed.

In an infinitesimal departure from the equilibrium state in which all the quantities U_i, V_i, and N_i change by infinitesimal amounts from their equilibrium values,

$$dS = dS_A + dS_B = 0 \qquad\qquad 11.20$$

Applying Equation 11.1 to the two phases A and B,

$$dS_A = \frac{1}{T_A}\left(dU_A + P_A dV_A - \mu_A dN_A\right) \qquad\qquad 11.21$$

$$dS_B = \frac{1}{T_B}\left(dU_B + P_B dV_B - \mu_B dN_B\right) \qquad\qquad 11.22$$

With these explicit values for dS_A and dS_B, Equation 11.20 becomes

$$\frac{1}{T_A}\left(dU_A + P_A dV_A - \mu_A dN_A\right) + \frac{1}{T_B}\left(dU_B + P_B dV_B - \mu_B dN_B\right) = 0$$

This expression can be simplified using the conservation conditions, from which $dV_A = -dV_B$, $dU_A = -dU_B$, and $dN_A = -dN_B$. Therefore

$$\left(\frac{1}{T_A} - \frac{1}{T_B}\right)dU_A + \left(\frac{P_A}{T_A} - \frac{P_B}{T_B}\right)dV_A - \left(\frac{\mu_A}{T_A} - \frac{\mu_B}{T_B}\right)dN_A = 0 \qquad 11.23$$

This must be true for <u>any</u> dU_A, dV_A, and dN_A. Therefore, Equation 11.23 is valid only if each of the factors in parentheses is identically equal to zero. Thus:

$T_A = T_B$ (the condition for thermal equilibrium).

$P_A = P_B$ (the condition for mechanical equilibrium).

$\mu_A = \mu_B$ (the condition for equilibrium against particle exchange).

This leads to a general conclusion:

> **If two phases or systems are in thermal and mechanical equilibrium, then they will also be in equilibrium against particle flow, and therefore the two phases or systems are in complete equilibrium if the chemical potentials are equal.**

Just as heat flows from regions of high to low temperatures until the temperatures become equal, particles flow from regions of high to low chemical potential until the chemical potentials become equal. The following example explains why this is so. Suppose there is a state very close to the final equilibrium state but with a small positive excess δN_A of particles in volume A over the equilibrium value N_A. There will be a corresponding particle deficit equal to $-\delta N_A$ in volume B. The entropy change in returning to the equilibrium state of maximum entropy must be slightly positive and is given by the sum of the last terms on the right of Equations 11.21 and 11.22. Therefore

$$\delta S = -\mu_A \left(\frac{dN_A}{T} \right) - \mu_B \left(\frac{dN_B}{T} \right) > 0$$

Note that the change in the population of A in returning to the equilibrium state is $-\delta N_A$ and that of B is $+\delta N_A$. This is done with no change in the temperature. Hence

$$\delta S = -\mu_A \left(-\frac{\delta N_A}{T} \right) - \mu_B \left(+\frac{\delta N_A}{T} \right) > 0 \qquad 11.24$$

or

$$\left(\mu_A - \mu_B \right) \delta N_A > 0 \qquad 11.25$$

Because δN_A is positive, Equation 11.25 shows that $\mu_A > \mu_B$, which means that the particle flow is from the region of high chemical potential to the region of low chemical potential.

The condition that chemical potentials of phases A and B are equal in equilibrium is related to the condition that the Gibbs functions per unit mass are equal in equilibrium at fixed T and P (Equation 10.3). This means that the Gibbs functions per molecule are also equal for the two phases, since they are composed of identical molecules. Further, because each phase is composed of only a single type of molecule or particle type, it follows from Equation 11.11 that the Gibbs function per molecule is actually the chemical potential for each phase. Thus, for change of phase, the equality of the specific Gibbs functions is equivalent to an equality of the chemical potentials for equilibrium against particle flow. This should not be surprising, because both results were derived from the same idea: the principle of increasing entropy.

The discussion in this section has involved the important case in which the boundary between the two systems is allowed to move. The simpler problem of the equilibrium conditions for two systems separated by a fixed permeable diathermal wall is left as Problem 11.2. Not surprisingly, the result is again equality of the chemical potentials.

11.2 THREE APPLICATIONS OF THE CHEMICAL POTENTIAL

The flow of matter from one system to another occurs so frequently in nature that the concept of chemical potential has applications in many fields. In this section, applications from the fields of biology, solid state physics, and chemistry are presented.

11.2.1 Osmotic Pressure

As a logical extension of the presentation in Section 11.1, it is natural to consider what happens when there are two or more different particle types. As an example, Figure 11.3 shows regions A and B separated by a rigid, diathermal wall that can sustain a pressure difference. Initially, let A contain a gas consisting of particles of type 1 only, and B a gas of particles of type 2 only (Figure 11.3(a)). Let the wall be permeable to gas 1 only. Such a wall, called

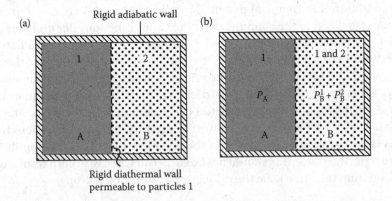

Figure 11.3 Semipermeable membrane separating the two regions A and B of the chamber is permeable to particles 1 only. This sets up an osmotic pressure across the membrane.

a *semipermeable membrane*, does occur in nature, particularly in biological systems. Such a membrane functions because it contains holes small enough to let through only particles smaller than a certain size. Although biological membranes are not strictly rigid, they may be modeled as rigid because they can sustain a pressure difference after an initial deformation.

After particles of gas 1 pass from A through the semipermeable membrane to B, the equilibrium state has a mixture of gases in B but only type 1 in A, as shown in Figure 11.3(b). The presence of additional particles of type 1 on the B side results in pressure $P_B{}^1$ on that side. The particles of type 2 remain on the B side and create pressure $P_B{}^2$, so the net pressure on that side is $P_B = P_B{}^1 + P_B{}^2$. It is straightforward to show that $P_B > P_A$. First, note that the chemical potentials for particles of type 1 are equal on the two sides in equilibrium. That is, $\mu_A{}^1 = \mu_B{}^1$ (see Problem 11.4). Now suppose that there is no interaction between particles of type 1 and type 2 in B. Given this assumption, the behavior of particle type 1 in B is the same as if the type 2 particles were absent. The equality of chemical potentials $\mu_A{}^1 = \mu_B{}^1$ means that the particles of type 1 exert the same pressure on both sides, provided they exist in the same phase (see Problem 11.3). In other words, $P_A{}^1 = P_B{}^1$. This completes the proof that $P_B = P_B{}^1 + P_B{}^2 > P_A$. It is this pressure difference that is known as *osmotic pressure*. Osmosis is of vital importance in the functioning of living organisms. In a cell, the osmotic pressure is balanced by the stresses in the cell wall. See for example the biophysics textbook by Cotterill (2002).

Although this discussion was framed in terms of gases, species 1 and 2 may be in the liquid state in biological systems, or as is usually the case, a solute dissolved in the liquid acting as a solvent. This affects the assumption that particles 1 and 2 are non-interacting. For liquids this assumption is not valid, but it acts as a first approximation. Correction terms must be added to the osmotic pressure to account for the interactions.

11.2.2 The Fermi Level

In the free electron theory of metals, electrons are treated as a gas of particles that obey quantum statistics. (See Chapter 13 for an introduction to quantum statistics.) Electrons are fermions, and according to the Pauli exclusion principle no two fermions may occupy the same quantum state. As a consequence, at absolute zero the electrons fill the lowest available quantum states up to a maximum energy called the *Fermi energy*. At higher temperatures this maximum energy often differs only very slightly from the Fermi energy. Any additional electron added to the metal has to be added at or minimally above the Fermi level, because all the lower energy levels are already occupied. This is

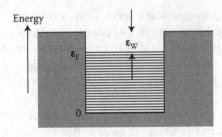

Figure 11.4   Schematic representation of the energy levels of the electrons in a metal.

represented schematically in the energy level diagram given in Figure 11.4. The *work function* ε_w is the energy required to liberate an electron from the metal at the Fermi level.

According to Equation 11.2, the chemical potential μ is the increase in the internal energy upon the addition of one particle, under conditions of constant entropy and volume. These conditions mean that the system is isolated from the surroundings apart from the reversible addition of the particle. Constant volume V implies that no work is done on the system, and the constant entropy S implies that no heat is added. When an electron is added to the metal, it enters at the Fermi level, and so the energy of the metal is then increased by ε_F. Therefore, μ can be identified with ε_F.

Now consider two different metals in contact. Each one has its own Fermi energy and work function. When they are put in contact so that there can be electron flow between them, the chemical potentials must be equal at equilibrium. This means that the Fermi levels are the same, and the energy level diagram is as shown in Figure 11.5. It can be seen from this that there will be a contact potential difference between them equal to $(\varepsilon_{W_2} - \varepsilon_{W_1})/e$ where e is the electron charge. This idea can be extended to semiconductors and is of fundamental importance in the operation of junction diodes, because the size of the contact potential affects the diode's operation.

11.2.3 The Condition for Chemical Equilibrium

Chemical potential is aptly named because of its particular importance in chemistry. As just one example of its use there, it is possible to derive a general condition for chemical equilibrium in a reaction.

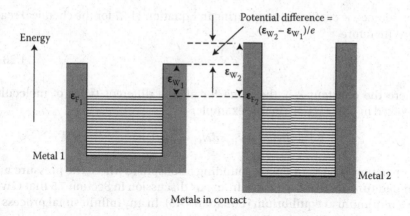

Figure 11.5   Physical origin of the contact potential. When two metals are put into electrical contact, their chemical potentials are equal at equilibrium. This means that their Fermi levels are the same, as indicated in the figure. The contact potential difference is equal to $\left(\varepsilon_{W_2} - \varepsilon_{W_1}\right)$, the difference in work functions divided by the electron charge.

For example, consider the common chemical reaction $H_2 + Cl_2 \rightarrow 2HCl$, which can be written as a mathematical equation

$$2HCl - H_2 - Cl_2 = 0 \qquad\qquad 11.26$$

A general way of writing such a reaction is

$$\sum_i v_i A_i = 0 \qquad\qquad 11.27$$

where the A_i denotes chemical symbols, and the so-called *stoichiometric coefficients* v_i are either positive or negative small integers. Assuming that the reaction proceeds in a given direction, then v_i is positive if its molecule is formed in the reaction and is negative if one disappears as a reactant, as in the example above. In the formation of HCl, $v_{H_2} = v_{Cl_2} = -1$ and $v_{HCl} = 2$.

Let N_i be the number of molecules of type i involved in the reaction. The numbers N_i will change as the reaction proceeds, but they cannot change independently of each other. The numbers can change only in a way that is consistent with the equation denoting the chemical reaction, because the numbers of the different types of atoms are conserved. For each molecule of H_2 and each molecule of Cl_2 that disappear upon reaction, two new molecules of HCl appear. The change in the number N_i must therefore be proportional to the stoichiometric coefficients appearing in Equation 11.27 for the chemical reaction. Therefore

$$dN_i = \lambda v_i \qquad\qquad 11.28$$

where the constant λ is the same for all the different types of molecules involved in the reaction. In this example,

$$dN_{HCl} : dN_{H_2} : dN_{Cl_2} = 2 : -1 : -1$$

If a reaction is open to the surrounding atmosphere where the pressure and temperature are fixed, we know from our discussion in Section 7.5 that G will be a minimum at equilibrium (Equation 7.45). In any infinitesimal process at equilibrium $dG = 0$, and so Equation 11.16 gives

$$dG = \sum_i \mu_i dN_i = 0 \qquad\qquad 11.29$$

Using Equation 11.28, this becomes

$$\sum_i \mu_i v_i = 0 \qquad\qquad 11.30$$

This is the general condition for chemical equilibrium in a reaction. It says that chemical potentials are additive, weighted by their stoichiometric coefficients. For example, in the reaction for the formation of HCl, Equation 11.30 gives

$$2\mu_{HCl} - \mu_{Cl_2} - \mu_{H_2} = 0$$

or

$$\mu_{HCl} = \frac{1}{2}\left(\mu_{Cl_2} + \mu_{H_2}\right)$$

which is a useful result.

EXAMPLE 11.1

The compound magnesium oxide is used to treat common digestive ailments. It is formed in the reaction

$$2Mg + O_2 \rightarrow 2MgO$$

Find the chemical potential for magnesium oxide as a function of the chemical potentials of the two reactants.

Solution: As described in the text, it is useful to write the reaction as an equation of the form $2Mg + O_2 - 2MgO = 0$ or $2Mg + O_2 = 2MgO$. From this form, it is evident that

$$\mu_{MgO} = \mu_{Mg} + \tfrac{1}{2}\mu_{O_2}$$

Problems

11.1 Derive Equation 11.3. [Hint: Express $U = U(S, V, N_1, N_2 \ldots)$ and use this to find an expression for dU.]

11.2 Consider two systems, A and B, each composed of the same single particle type. The two systems are contained in a chamber surrounded by rigid adiabatic walls and are separated from each other within the chamber by a rigid diathermal wall that is also permeable to the particles (Figure 11.6). Show, using an argument similar to the one used in Section 11.1.4, that the condition for equilibrium against particle exchange is the equality of the chemical potentials.

11.3 Consider the system of Problem 11.2. Suppose that the two systems are composed of the same single type of particle and are both in the same phase, for example a gas on each side of the separating wall. Show that the pressures are equal. Would the pressures be equal if different phases existed on either side of the wall? [Hint: For a system consisting of a single type of particle, $\mu = G(T, P, N)/N = \phi(T, P)$. If the phases are the same on either side of the wall, the function ϕ must also be the same on either side.]

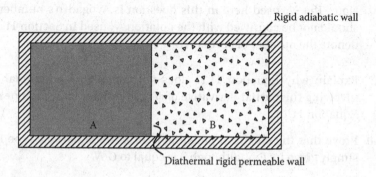

Rigid adiabatic wall

A

B

Diathermal rigid permeable wall

Figure 11.6

11.4 Consider the system of Problem 11.2, but now suppose there is a mixture of gases on either side of the separating wall that is permeable to all the different gases i. Argue that $\mu_A{}^i = \mu_B{}^i$ for all the different gases i. [Hint: Place another diathermal wall, permeable to only one gas, in front of the separating wall and proceed as in Section 11.1.4.]

11.5 Show that the chemical potential of an ideal gas at temperature T varies with pressure as

$$\mu = k_B T \ln\left(\frac{P}{P_0}\right) + \mu_0$$

where μ_0 is the value at reference point of pressure P_0 and temperature T. The gas consists of a single type of particle only. This expression is of great use in chemistry. [Hint: $V = (\partial G/\partial P)_{T,N}.$]

11.6 Show that the chemical potential of an ideal monatomic gas of N particles is

$$\mu = \frac{5}{2}k_B T - \frac{3}{2}k_B T \ln T - k_B T \ln\left(\frac{N_A V}{N}\right) - \frac{S_0 T}{N_A}$$

Hints:

i. Use Equation 5.13 for S.

ii. $F = U - TS$ and $\mu = (\partial F/\partial N)_{T,V}$ or $G = U + PV - TS$ and $\mu = G/N$.

iii. $U = 3Nk_B T/2$.

iv. $C_V = 3Nk_B/2$.

Note: The N_A used here in this question is Avogadro's number and should not be confused with the notation N_A used in Section 11.1.4 to denote the number of particles in a box A.

11.7 Explain why Equation 11.11 does not contain an additional term involving the partial derivative of ϕ with respect to N, as appears in Equation 11.13.

11.8 Prove that the chemical potential μ for a single particle type is not simply given by F/N, although it is equal to G/N.

11.9 Helium gas at $T = 298$ K and $P = 1$ atmosphere has chemical potential -0.32 eV. (a) Explain the significance of the negative sign. (b) How much pressure would need to be applied (at constant T) for the chemical potential to be exactly zero? [Hint: Use the result of either Problem 11.5 or Problem 11.6.]

11.10 Ammonia NH_3 is formed by combining nitrogen N_2 and hydrogen H_2. Find the chemical potential for ammonia as a function of the chemical potentials of the two reactants.

11.11 The Sackur–Tetrode equation (similar to Equation 5.11) gives the entropy of an ideal gas as

$$S = Nk_B \left\{ \ln \left[\frac{V}{N} \left(\frac{4\pi mU}{3Nh^2} \right)^{3/2} \right] + \frac{5}{2} \right\}$$

where h is Planck's constant. (a) Show that the chemical potential can be written in terms of entropy as

$$\mu = -T \frac{\partial S}{\partial N} \bigg|_{U,V}$$

(b) Use the result of part (a) along with the fact that $U = (3/2)Nk_BT$ for a monatomic gas to find an expression for the chemical potential as a function of V, N, and T. (c) Evaluate the result in (b) numerically for helium gas at $T = 298$ K and $P = 1$ atm. (d) Discuss the implications of the fact that your answer in (c) is negative.

REFERENCE

Cotterill, R., *Biophysics: An Introduction*, Wiley, West Sussex, 2002.

Chapter 12: The Third Law of Thermodynamics

The third law of thermodynamics is often stated as: The entropy of a system must approach zero in the limit of zero temperature. This statement may sound simple, even obvious, but there are several alternative ways to think of the third law, and as a result, some interesting consequences emerge after deeper investigation.

12.1 STATEMENTS OF THE THIRD LAW

12.1.1 The Nernst Heat Theorem

The third law of thermodynamics is concerned with the entropy of a system as the temperature is reduced toward absolute zero. Integrating Equation 5.4 from absolute zero to temperature T,

$$S = \int_0^T \frac{dQ}{T} + S_0 \qquad 12.1$$

In this case S_0, the entropy at absolute zero, cannot be determined from the second law. This is where the third law becomes meaningful, providing a value for S_0.

The original statement of the third law was given by Walther Nernst in 1906. Nernst noticed that, in many chemical reactions occurring with no change in the end point temperatures, the value of ΔG decreased while that of ΔH increased. Nernst postulated as his heat theorem that not only do these two quantities become equal at $T = 0$, but they also approach each other asymptotically, as shown in Figure 12.1.

DOI: 10.1201/9781003299479-12

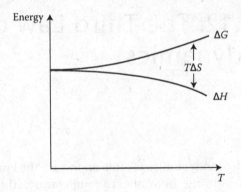

Figure 12.1 Nernst postulated that the curves for ΔH and ΔG in a chemical reaction approach each other asymptotically as $T \to 0$.

Consider what this means in terms of entropy. It follows from Equation 7.37 that

$$\Delta G = \Delta H - T\Delta S \qquad\qquad 12.2$$

for a chemical reaction at temperature T, and this clearly shows that $\Delta G \to \Delta H$ as $T \to 0$. However, in order for the curves to touch each other asymptotically as in Figure 12.1, it can be shown that ΔS itself must vanish as $T \to 0$. These ideas are embodied in the following statement of the Nernst heat theorem, which is more general than the one originally given by Nernst:

> **The entropy change in a process that occurs between a pair of equilibrium states, associated with a change in an external parameter, tends to zero as the temperature approaches absolute zero.**

The external parameters may be, for example, the pressure, temperature, or magnetic field. Ever since its formulation, there has been discussion about the significance of the Nernst heat theorem. There is now a great deal of experimental evidence in its support, and so it has assumed the status of a fundamental law—the third law of thermodynamics.

12.1.2 Planck Formulation of the Third Law

The original Nernst formulation of the third law has subsequently been followed by various other statements that are more in accord with an understanding of modern quantum statistical mechanics. A particularly useful statement of the third law was given by Planck in 1911. It is a more powerful statement than the earlier statement of Nernst and is

> **The entropy of all perfect crystals is the same at absolute zero and may be taken as zero.**

A perfect crystal is one in which the arrangement of the atoms repeats itself on a regular basis throughout the crystal, with no imperfections. In a glass or amorphous material, on the other hand, there is no such regular repetition and no long-range order. The essential point in the Planck statement is that the entropies of perfect crystals are equal at $T = 0$; it is then a matter of convenience to put $S_0 = 0$. This is a sensible choice, because it gives agreement with the microscopic view to be discussed in Section 12.1.4. Although the Planck statement is usually quoted for perfect crystals for historical reasons, it is now believed to hold for all systems that are in equilibrium states, including liquids (e.g., ^{3}He and ^{4}He) and gases.

> **It might seem surprising that one can consider a gas existing at absolute zero. However, there are certain quantum systems, such as the electrons in a metal (Section 13.3), which do constitute a gas-like assembly even down to $T = 0$. Also, extremely dilute gases such as those in a Bose–Einstein (BE) condensate (Section 13.4) exist at nanokelvin temperatures.**

In this context, it is important to understand what is meant by an equilibrium state, from an energy point of view. It will also be important to consider the difference between *stable* and *metastable* equilibrium. Under given conditions, a system in equilibrium is in a state corresponding to a minimum in the appropriate potential function. For example, a complex system at constant P and T takes a minimum value of G at equilibrium when another macroscopic variable, such as the volume, is varied. Such a state is one of stable equilibrium because any departure from this state entails an increase in G. On the other hand, if G increases for small departures only from the equilibrium state before decreasing again for larger departures, then the state is one of metastable equilibrium.

12.1.3 Experimental Test of the Third Law

The ideas presented in Section 12.1.2 allow a laboratory test of the validity of the third law. Many substances exist in different allotropic forms. Here are two examples:

1. Pure tin exists in two forms, called gray and white tin. Above 286 K, the stable form is white tin (also called β-tin), which is metallic with a tetragonal crystal structure. Below that temperature, the stable form is gray tin (also called α-tin), which is powdery and has a face-centered diamond-cubic structure.

2. Sulfur has many allotropes, but at atmospheric pressure, the common ones are monoclinic and rhombohedral sulfur. Above 368 K, the stable form is monoclinic (β-sulfur), while below this temperature, the stable form is rhombohedral (α-sulfur).

> There are two other allotropes of tin, called γ- and σ-tin, but these exist only at high temperatures and pressures greater than 10^4 atm.

Consider now one such substance that can exist in two allotropic ordered forms. At a given temperature, each form can exist in an equilibrium state with a minimum in energy, although the more stable form will have the lower energy minimum. The form with the higher energy minimum is in metastable equilibrium. If sufficient energy is given to the system in this metastable equilibrium state, so that the potential barrier between the states can be overcome, the system will change to the lower energy state of stable equilibrium. This is analogous to a ball being kicked out of a small depression in a hillside so that it rolls down to the bottom of a hill. At high temperatures, where $k_B T$ is larger than or comparable to the height of the potential barrier, there is sufficient random thermal vibrational energy to induce such a change; thus a system initially in the metastable state will change gradually into the stable state. The metastable state at high temperatures is not in a true equilibrium state, because the state changes with time. However, at low temperatures, and certainly at absolute zero, the situation is different. Now, such a metastable state is in a true equilibrium state, because there is insufficient thermal energy to induce a change. As a result, the Planck form of the third law may be applied to the system in either the stable or the metastable equilibrium states.

As a specific application of these ideas, consider the allotropes of sulfur described above. If monoclinic sulfur is cooled rapidly through 368 K to a very low temperature, the monoclinic phase is locked in, with the transition rate to the more stable rhombohedral form becoming negligible. Thus, at low temperatures both forms of sulfur can be produced in stable equilibrium states, with S_0 being the same for both forms. If the Planck statement of the third law holds, then the entropy must vary for the two forms as in Figure 12.2, with a

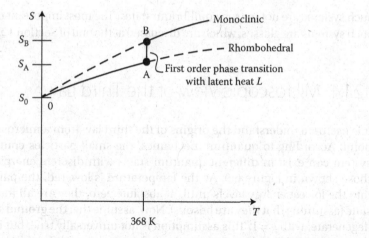

Figure 12.2 The entropies of monoclinic and rhombohedral sulfur as a function of temperature. The curves may be obtained from a measurement of the temperature dependence of the heat capacity for each phase. Based on these experimental measurements, together with a measurement of the latent heat for the first-order phase transition at 368 K, the entropy curves must meet each other at $T = 0$. This is consistent with the third law.

common value of S_0. The entropy change $S_B - S_0$ in monoclinic sulfur between 0 and 368 K is given by the two expressions

$$S_B - S_0 = \int_0^{368} \frac{C_P^{\text{monoclinic}}}{T} dT$$

and

$$S_B - S_0 = S_B - S_A + S_A - S_0 = \frac{L}{368} + \int_0^{368} \frac{C_P^{\text{rhombohedral}}}{T} dT$$

where L is the latent heat for the transition (Section 3.3.4). Now, the heat capacities for each of the two forms can be measured from close to absolute zero up to 368 K. Also, L can be measured, and so the two values for $\dot{S}_B - S_0$ can be determined. Typical experimental values for one mole are $s_B - s_0 = 37.82 \pm 0.40$ and 37.95 ± 0.20 kJ/(K·mol). Because these values are the same within the experimental error, the assumption that S_0 is the same for each form of sulfur is valid. This is also found for other different allotropes, which supports the Planck statement of the third law.

Although there are some systems that appear to violate the third law, in that they have non-zero entropy at absolute zero, there is in fact never any violation because such systems are not in true equilibrium states. The most important examples of such systems are glasses, which are discussed at the end of Section 12.1.5.

12.1.4 Microscopic View of the Third Law

It is useful to understand the origins of the third law from a microscopic viewpoint. According to quantum mechanics, the small particles comprising the system can exist in different quantum states with discrete energies, such as those shown in Figure 6.9. As the temperature is lowered, the particles pack into the lowest energy levels until, at absolute zero, they are all in the ground state (assuming that they are bosons). Now, assume that the ground state is non-degenerate, with $g = 1$. This assumption is not universally true, but it simplifies the argument. With these assumptions, the number of different ways in which all the particles can be arranged in the ground state is $\Omega = 1$. An application of

$$S = k_B \ln \Omega \qquad\qquad 6.3$$

shows that $S_0 = 0$ at absolute zero, which is the Planck statement of the third law. This of course also implies that $\Delta S = 0$ for any process occurring at absolute zero, which is precisely the Nernst heat theorem.

This argument for the approach of entropy to zero at absolute zero hinged on the discreteness of the energy levels, as well as on the non-degeneracy of the ground state. However, for the macroscopic systems of concern in thermodynamics, these levels are very closely spaced. At temperatures of a few degrees K, where the measured decrease in S from the room temperature value is quite marked, there are so many states that may be occupied that one cannot say the system is in one lowest energy state with $\Omega = 1$. The explanation is that the decrease in entropy toward zero depends on the behavior of the number of states per unit energy range (called the *density of states*) with energy, rather than just on the occupancy of the ground level, as this simplified theory suggested. In particular, it is the behavior of the density of states at low energies that determines the low temperature properties, so the entropy certainly does fall to zero as T tends to 0 for both bosons and fermions. The density of states functions for bosons and fermions will be described in Chapter 13.

12.1.5 Simon Formulation of the Third Law

Consider cooling a perfect crystal (described in Section 12.1.2) toward absolute zero. As the crystal is cooled, lattice vibrations decrease and a state of perfect order is approached, with the atoms all settling in their lattice positions and S tending toward zero. However, the electrons in the atoms can have a net electron magnetic moment due to spin. If the temperature is merely very low but not zero, the electrons can have different energies according to whether the spins associated with their magnetic moments point along or against an applied magnetic field. There will thus be some entropy remaining, due to the different orientations of the electron spins, when all the lattice vibrational entropy has disappeared.

Now, cool the crystal even more, so that the electron spins all go into the same ground state, aligned with the magnetic field. This is the most ordered arrangement, with zero spin entropy. Is the system's entropy equal to zero? The answer is almost certainly no, because one must now consider the weak nuclear spin magnetic moments, which again can be distributed among a set of very closely spaced energy levels. The temperature must be reduced even further to remove this nuclear spin entropy. Of course, this argument could be continued by considering other contributions to the entropy due to the arrangements of individual nucleons within each nucleus. In the second statement of the third law (Section 12.1.2), these hidden entropies were disregarded by stipulating a perfect (and by implication simple) crystal in which there are no electron or nuclear spins, only atomic size masses in a regular arrangement.

The important point is that the lattice, the electron spin system, and the nuclear spin system are essentially uncoupled from each other at low enough temperatures. Then, they act as independent systems, each in internal thermodynamic equilibrium. For such uncoupled systems, the entropies are additive. Simon called these independent systems *aspects* of the whole system and gave the following general statement of the third law in 1937:

> **The contribution to the entropy of a system from each aspect that is in internal thermodynamic equilibrium disappears at absolute zero.**

The Simon statement is convenient, because it means that attention can be focused on just one aspect of interest, with the knowledge that its entropy is zero at $T = 0$.

Finally, consider what happens in glasses, which have no crystal structure whatsoever. Many liquids, which by nature have high entropy, retain their liquid structure if they are cooled rapidly through their freezing point to form a glass, with frozen-in entropy at absolute zero. In contrast, if they are cooled slowly, they first go into an ordered crystal phase, with zero entropy at absolute zero. The crystal phase is the stable phase, with a minimum in energy, while the glass is an unstable phase, not at an energy minimum. The glass will slowly crystallize, although the time period for this may be years or even considerably longer. The glass phase is thus not an equilibrium state. Glycerol, with a melting point at 18°C, is a good example of such a material. Although the frozen-in entropy persists down to absolute zero, this in no way violates the third law, which relates only to systems in equilibrium.

12.2 CONSEQUENCES OF THE THIRD LAW

It is a consequence of the third law that not only entropy but also several other measurable parameters vanish at absolute zero.

12.2.1 Thermal Expansion Coefficient

The thermal expansion coefficient β is defined as

$$\beta = \frac{1}{V}\left(\frac{\partial V}{\partial T}\right)_P \qquad 2.2$$

This can be transformed by the Maxwell relation

$$\left(\frac{\partial V}{\partial T}\right)_P = -\left(\frac{\partial S}{\partial P}\right)_T \qquad 7.41$$

to

$$\beta = -\frac{1}{V}\left(\frac{\partial S}{\partial P}\right)_T \qquad 12.3$$

• However, the Nernst formulation of the third law says that the entropy change in an isothermal process tends to zero at absolute zero, so the partial derivative in Equation 12.3 is zero. Therefore, the thermal expansion coefficient is zero at absolute zero.

The practical effect of this is that very small changes in temperature near absolute zero result in negligible thermal expansion. Considering the discussion in

Section 12.1.5, it is likely that a small amount of thermal energy will go into other forms of energy, rather than expanding the crystal lattice.

12.2.2 Temperature Dependence of the Magnetic Moment in a Magnetic System

In Section 9.1.4, it was suggested how to derive a Maxwell relation for magnetic systems

$$\left(\frac{\partial S}{\partial B_0}\right)_T = \left(\frac{\partial M}{\partial T}\right)_{B_0} \qquad 12.4$$

As was the case for thermal expansion in the preceding section, the left side of Equation 12.4 vanishes in the limit as T approaches zero, by the Nernst formulation of the third law. Hence, $(\partial M/\partial T)_{B_0} = 0$ at absolute zero, which means that there is no temperature dependence of the magnetic moment at $T = 0$. This immediately suggests that the Curie law cannot hold down to absolute zero, as the following argument shows.

Suppose the Curie law is obeyed as

$$\chi_m = \frac{C}{T} = \frac{\mu_0 M}{V B_0} \qquad 12.5$$

where the magnetic susceptibility follows from Equation 9.5. Differentiating,

$$\left(\frac{\partial M}{\partial T}\right)_{B_0} = -\frac{V B_0 C}{\mu_0 T^2} \qquad 12.6$$

Clearly, the temperature dependence of the magnetic moment does not vanish as $T \to 0$. Therefore, the conclusion is that the Curie law breaks down at very low temperatures.

The physical reason for the breakdown of Curie's law in a magnetic system is that there is always an interaction between the elementary magnetic dipoles, whereas the Curie theory of paramagnetism assumes only interaction with the external field. In a magnetic salt, this interaction is usually very weak and is negligible compared with the thermal energy $k_B T$ at normal or even fairly low temperatures; as a result, Curie's law holds. At very low temperatures, the interaction between dipoles becomes important, so the salt departs from Curie's law. In that case, some sort of magnetic ordering occurs, and the salt can become ferromagnetic or antiferromagnetic. The magnetic moment then does not change with temperature at $T = 0$, as was deduced from Equation 12.4.

12.2.3 Heat Capacities

Next, consider the constant-volume heat capacity C_V. It is related to entropy and temperature by

$$C_V = T \left(\frac{\partial S}{\partial T} \right)_V \qquad 7.9$$

or

$$C_V = \left(\frac{\partial S}{\partial \ln T} \right)_V \qquad 12.7$$

as $T \to 0$, $\Delta S \to 0$ by the Nernst statement of the third law. But $\Delta(\ln T)$ is certainly non-zero as $T \to 0$, because $\ln T \to -\infty$ there. Hence, heat capacity C_V should tend to zero as $T \to 0$. A similar argument holds for the other heat capacities.

This is always observed experimentally. For example, the heat capacity of most metals at low temperatures is found to obey the law

$$C_P = aT + bT^3 \qquad 12.8$$

(where a and b are constants for a particular material) down to the very lowest temperatures attainable. The first term is due to the conduction electrons, and the second is the contribution from the lattice. This heat capacity vanishes at $T = 0$, in agreement with the third law.

There is a more instructive argument for the vanishing heat capacities at absolute zero. From Equation 7.9, the entropy change from absolute zero to temperature T is

$$S - S_0 = \int_0^T \frac{C_V}{T} dT \qquad 12.9$$

The left side of this equation must remain finite at all temperatures down to absolute zero, where it vanishes by the third law. Therefore, the right side of Equation 12.9 must also remain finite. This means that C_V must fall to zero as

$T \rightarrow 0$, at least as fast as T; otherwise, the right side would diverge there. Equation 12.8 is consistent with this idea.

12.2.4 Slope of the Phase Boundary in a First-Order Transition

The slope of the phase boundary in a first-order transition is given by the Clausius–Clapeyron equation

$$\frac{dP}{dT} = \frac{\Delta S}{\Delta V}$$

10.8

As $T \rightarrow 0$, ΔS tends to zero by the Nernst statement of the third law. Therefore, the slope of a phase boundary in a first-order transition is zero at absolute zero. This is shown in Figures 10.13 and 10.14, the phase diagrams for ^{4}He and ^{3}He.

Another application of this idea is in superconductors. It can be shown in exactly the same way that the phase boundary between the normal and the superconducting phases has zero slope at $T = 0$. This is illustrated in Figure 10.11.

> **Based on all the examples in Section 12.2, the fact that all the predictions based on the third law are in agreement with experimental observations may be taken as the experimental confirmation of the third law.**

12.3 THE UNATTAINABILITY OF ABSOLUTE ZERO

There is yet another statement of the third law:

> **It is impossible to reach absolute zero using a finite number of processes.**

Although there are formal proofs to show that this statement of the third law is equivalent to the Nernst statement (see, e.g., the text by Zemansky 2013), we shall resort to a physical argument to show that the two statements are consistent with each other.

The lowest temperatures attainable experimentally in bulk materials are achieved using the technique of adiabatic demagnetization (Section 9.2). One might logically propose to perform a series of successive demagnetizations in an attempt to reach absolute zero, as illustrated in Figure 12.3.

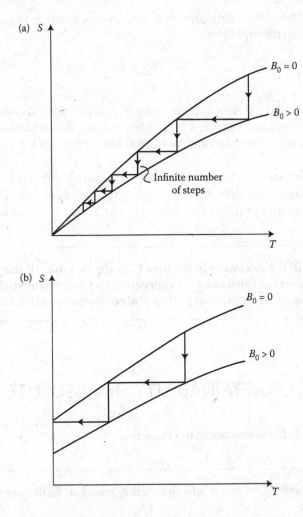

Figure 12.3   It is impossible to reach absolute zero in a finite number of steps. The process illustrated here is for the adiabatic demagnetization of a paramagnetic salt. (a) If the third law is true, the approach toward absolute zero is gradual but is never quite achieved. (b) If a third law were untrue, absolute zero could be reached after a finite number of steps.

Figure 12.3 shows two entropy curves for a magnetic salt. One curve represents a finite magnetic field B_0, and the other shows the entropy for zero applied field. As the elementary dipoles are aligned by the field, they are in a state of greater order than when they are unaligned in the absence of the external field. Thus, the entropy curve for the state in the applied field is the lower one. At absolute zero, by the Nernst statement of the third law, the difference in entropy between the two curves is zero, and this is shown in Figure 12.3(a). In contrast, Figure 12.3(b) shows the two entropy curves in a way that violates the third law.

Now, consider the series of isothermal magnetizations and adiabatic demagnetizations represented by the zigzag paths in Figure 12.3. Each successive demagnetization reduces the temperature. If the entropy curves were as in Figure 12.3(b), then absolute zero could be reached in a finite number of operations. However, the actual entropy curves are as in Figure 12.3(a), and it is clear that absolute zero cannot be obtained in a finite number of demagnetizations. Therefore, it is impossible to reach absolute zero in any practical way.

Despite the unattainability of absolute zero, much progress has been made in recent years in reaching extremely low temperatures. In Section 10.5.2, the discovery of the superfluid phases of ^{3}He was described, at the extremely low temperatures of a few millikelvin. Since then, laboratories have routinely achieved lattice temperatures of a few microkelvin, while the nuclei of copper have been cooled to a few nanokelvin. At these temperatures, they form a nuclear antiferromagnet. The Bose–Einstein condensates (BECs) described in Section 13.4 also occur at nanokelvin temperatures, with the aid of laser cooling of atomic gases. Absolute zero will always be unattainable, but a wealth of fascinating physics remains to be discovered as laboratory techniques improve, allowing us to approach ever closer to this elusive end.

Problems

12.1 Discuss the behavior of the compressibility (defined in Equation 2.3) in the limit as the temperature approaches absolute zero. Does compressibility, like heat capacity and thermal expansion coefficient, approach zero in this limit? Explain.

12.2 Consider a mole of conduction electrons in copper, initially at absolute zero. Use the following process to estimate the temperature associated with promoting a single electron from the ground state to the first excited state. (a) Assume a single electron is promoted, and that the energy required is $\Delta\varepsilon$. This implies that $g(\varepsilon)\Delta\varepsilon = 1$, where $g(\varepsilon)$ is the density of states function (Equation 13.45). FD statistics require that the promoted electron originally has energy ε_F, the Fermi energy. Show that

$$\Delta\varepsilon = \frac{2\varepsilon_F}{3N}$$

where N is the number of electrons. (b) Use the Boltzmann factor from Chapter 6 to argue that the probability of an electron being in the first excited state, relative to the ground state, is

$$e^{-\Delta\varepsilon/k_B T}$$

(c) If the probability in (b) is small, show that the temperature must be given by the inequality

$$T < \frac{2\varepsilon_F}{3Nk_B}$$

(d) Evaluate the result in (c) for one mole of conduction electrons, with $\varepsilon_F = 7.0$ eV. Discuss the implications for the third law.

12.3 Explain how the process in Problem 12.2 would differ for a condensed boson gas. Would the maximum temperature be higher or lower for bosons than for fermions?

12.4 Using the kinetic theory of gases, the molar heat capacity of a monatomic ideal gas is $c_V = 3/2R$ (Section 3.4.3). Use the third law to discuss the suitability of kinetic theory in the limit of low temperatures.

BIBLIOGRAPHY

Baierlein, R., *Thermal Physics*, Cambridge University Press, Cambridge, 1999.

Reif, F., *Fundamentals of Statistical and Thermal Physics*, Waveland Press, Long Grove, IL, 2008.

Schroeder, D., *An Introduction to Thermal Physics*, Oxford University Press, Oxford, 2021.

Zemansky, M.W., *Heat and Thermodynamics*, McGraw-Hill, New York, 2013.

Chapter 13: Quantum Statistics

In Chapter 6, statistical principles were applied to collections of atoms or molecules in order to understand the thermodynamic behavior of larger systems. That approach was particularly useful for the study of gases, in which the behavior of individual particles can be understood with classical principles. There are many other systems, however, in which the quantum behavior of particles must be considered, and this requires the development of new statistical distributions appropriate for different kinds of particles. This leads to some unusual and fascinating phenomena, some of which were observed only recently.

13.1 CLASSICAL AND QUANTUM STATISTICS

In the classical gases studied in Chapter 6, the density of particles is so low that there is essentially no overlap between the particles' quantum mechanical wave functions. That is why quantum mechanics can be ignored. However, there are systems in which the density of particles is large enough that the classical approximation is no longer valid. Fortunately, the basic concepts of statistics such as multiplicity and probability are still useful when multiple particles are fighting for the same space, but the statistical rules for how the particles fill the space must be redeveloped.

13.1.1 Bosons and Fermions

In quantum mechanics, fundamental particles are described as *bosons* if they have zero or integer spin and *fermions* if they have half-integer spin. For example, leptons (such as electrons) and quarks are fermions with spin 1/2 and photons are bosons with spin 1. Composite particles, including mesons,

DOI: 10.1201/9781003299479-13

baryons, and whole atoms, are classified similarly based on their net spin. For example, the common ^{4}He atom is a boson, but the rare ^{3}He atom is a fermion. This apparently slight difference results in vastly different behavior in the superfluid phases of the two isotopes.

The important difference statistically is that fermions are governed by the *Pauli exclusion principle*, which says that no two fermions may occupy the same quantum state. A familiar example of this is the filling of ground-state energy levels in multielectron atoms. The lowest (in energy) electron subshell is the 1s subshell. Because electrons are fermions, the 1s subshell is restricted to two electrons, which are in different quantum states due to their opposite spins. A third electron cannot fit in the 1s subshell, because it would need to have the same spin (and same quantum state) as one of the other two. The same restriction holds for the 2s subshell. The 2p subshell has three different angular momentum states, so this combined with the spin degeneracy allows a total of six electrons in a p subshell. The other subshells fill similarly, which leads to the atomic properties described in the periodic table. Based on the Pauli exclusion principle, we say that fermions follow Fermi–Dirac (FD) statistics.

In contrast, there is no such restriction on bosons. In principle, any number of bosons may be in the same quantum state. As a result, bosons will be described by Bose–Einstein (BE) statistics.

For classical systems, there is no need to consider whether particles are fermions or bosons, because their wave functions are considered non-overlapping. For that reason, there is no restriction on how many classical particles may be in the same state, except for the phase-space restriction discussed in Section 6.2.5. Therefore, classical particles are described by a third type of statistics, called Maxwell–Boltzmann (MB) statistics. What distinguishes MB from BE is that classical particles (governed by MB statistics) can generally be considered distinguishable from one another, whereas quantum particles (whether fermions or bosons) must be considered indistinguishable.

13.1.2 Classical and Quantum Distributions: An Example

A simple example will illustrate some distinctions among the three distributions: MB, BE, and FD. Suppose a system consists of two particles, which have two states available to them. The possible distributions of the two particles between the two states is shown in Figure 13.1 for MB, BE, and FD statistics. For MB statistics, the particles are assumed distinguishable, and thus labeled A and B. For the two quantum distributions, the particles are indistinguishable, and so both are labeled A.

Even in such a simple system, the number of configurations depends on the kind of statistics used. The Pauli exclusion principle imposes a strong restriction on fermions, so the number of configurations is smallest for FD. The two fermions cannot be in the same state, and as a result, one fermion occupies each of the available states. Bosons are not subject to the Pauli exclusion principle. Therefore, as shown in Figure 13.1, the two bosons may be in different states or the same one. Thus, there are more configurations possible for BE than for FD. The difference between BE and MB is due to the distinguishability of classical particles.

Notice that for BE statistics, it is possible (even likely) that all the particles may cluster in the same state. This has important implications in some applications of BE statistics.

MB Statistics

S1	S2
A	B
B	A
AB	
	AB

FD Statistics

S1	S2
A	A

BE Statistics

S1	S2
A	A
AA	
	AA

Figure 13.1 Filling two states with two particles, using MB, FD, and BE statistics. FD statistics are the most restrictive, and MB statistics are the least.

13.1.3 Gibbs Factor and Grand Partition Function

Some important tools for dealing with quantum statistics may be derived by taking an approach analogous to the use of the Boltzmann factor and partition function in Chapter 6. In Section 6.3.1, the Boltzmann factor was developed by considering a system attached to a reservoir, assuming that the system and reservoir were fixed in size (volume and particle number) but could exchange energy.

When dealing with quantum statistics, it will be convenient to consider systems in which both particles and energy can be exchanged with the reservoir. The importance of this is illustrated by the example in Section 13.1.2. Allowing for particle exchange leads to a modification of the Boltzmann factor from Chapter 6. Recall that the thermodynamic identity (Equation 11.1) relates changes in entropy to both energy and particle number

$$dU = TdS - PdV + \mu dN \qquad\qquad 11.1$$

where μ is the chemical potential. This suggests that the energy E in the Boltzmann factor should be replaced by $E - \mu N$, to take into account particle flux, if we still assume an isochoric process. This modification results in a modified Boltzmann factor, now called the *Gibbs factor*, defined as

$$\text{Gibbs factor} = e^{-(E-\mu N)/k_B T} \qquad\qquad 13.1$$

Analogous to the partition function for classical statistics (Equation 6.11), the sum of all the Gibbs factors is defined as the *grand partition function* $\mathcal{Z}$:

$$\mathcal{Z} = \sum_i g_i e^{-(E_i - \mu N_i)/k_B T} \left(\text{grand partition function}\right) \qquad\qquad 13.2$$

In Equation 13.2, g_i is the degeneracy of a state with energy E_i and occupancy N_i. Then, as in classical statistics, the probability of a state with energy E_i is

$$P(E_i) = \frac{1}{\mathcal{Z}} g_i e^{-(E_i - \mu N_i)/k_B T} \qquad\qquad 13.3$$

13.1.4 FD Distribution

Consider a system made up of non-degenerate quantum states that have energy ε when occupied by a single particle. The net energy of n particles in such a state is $n\varepsilon$. The grand partition function is given by Equation 13.2 as

$$Z = \sum_n e^{-(n\varepsilon - \mu n)/k_B T} = \sum_n e^{-n(\varepsilon - \mu)/k_B T} \qquad 13.4$$

where, as in Equation 11.1, μ is the chemical potential. Similarly, the probability that a state is occupied by n particles is then given by Equation 13.3 as

$$P_n = \frac{1}{Z} e^{-(n\varepsilon - \mu n)/k_B T} = \frac{1}{Z} e^{-n(\varepsilon - \mu)/k_B T} \qquad 13.5$$

It is most useful to consider the mean number of particles $\bar{n}$ for the state, which is the weighted average

$$\bar{n} = \sum_n n P_n \qquad 13.6$$

In this context, *non-degenerate* means that there is only a single state corresponding to each energy ε.

Now, it becomes crucial to consider which kind of statistics apply. For FD statistics, each state can have either zero particles or one particle ($n = 0$ or $n = 1$). Therefore, for a given quantum state with energy ε, the grand partition function contains only two terms

$$Z = \sum_n e^{-n(\varepsilon - \mu)/k_B T} = 1 + e^{-(\varepsilon - \mu)/k_B T}$$

The weighted average $\bar{n}$ becomes (for fermions)

$$\bar{n}_{FD} = \sum_n n P_n = 0 + P_1 = \frac{e^{-(\varepsilon - \mu)/k_B T}}{1 + e^{-(\varepsilon - \mu)/k_B T}}$$

Simplifying,

FD distribution

$$\bar{n}_{FD} = \frac{1}{e^{(\varepsilon-\mu)/k_B T}+1}$$

13.7

This important result is known as the *FD distribution*. To understand its physical significance, consider the graphs shown in Figure 13.2. At $T = 0$, the graph is a step function with $\bar{n}_{FD} = 1$ for $\varepsilon < \mu$ and $\bar{n}_{FD} = 0$ for $\varepsilon > \mu$. This makes sense physically, because at $T = 0$ all the available states fill from lower energy to higher with one fermion per state, until every fermion is placed. When $T > 0$, thermal energy is available to promote some of the fermions to higher energies, and the distribution "smears" as shown.

13.1.5 BE Distribution

For bosons, there is no restriction on the number of particles per state, and therefore, the sums in Equations 13.4 and 13.6 each have an infinite number of terms. Fortunately, the sums can be computed using standard mathematical techniques.

Again, we will assume single-particle states with energy ε. The grand partition function (Equation 13.4) for bosons in a single state of energy ε is

$$\mathcal{Z} = \sum_n e^{-n(\varepsilon-\mu)/k_B T} = 1 + e^{-(\varepsilon-\mu)/k_B T} + e^{-2(\varepsilon-\mu)/k_B T} + \cdots$$

13.8

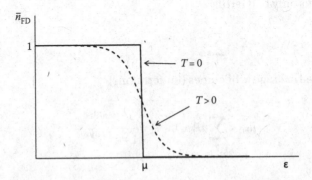

Figure 13.2 FD distribution at $T = 0$ and $T > 0$. At $T = 0$, all the levels are filled up to energy μ, and at higher temperatures, the shape of the distributions smears to allow some occupancy above μ.

This takes the form of a geometric series $1 + y + y^2 + \cdots$ with well-known result

$$1 + y + y^2 + y^3 + \cdots = \frac{1}{1-y}$$

In this context, $y = e^{-(\varepsilon-\mu)/k_B T}$, so the result is

$$Z = \frac{1}{1 - e^{-(\varepsilon-\mu)/k_B T}} \qquad 13.9$$

The weighted average occupancy (Equation 13.6) for bosons is also an infinite sum

$$\bar{n}_{BE} = \sum_n n P_n = \frac{1}{Z} \sum_n n e^{-n(\varepsilon-\mu)/k_B T} \qquad 13.10$$

This sum can be computed by letting $u = (\varepsilon-\mu)/k_B T$ and then realizing that

$$n e^{-nu} = -\frac{\partial}{\partial u} e^{-nu}$$

Therefore,

$$\bar{n}_{BE} = \frac{1}{Z} \sum_n n e^{-nu} = -\frac{1}{Z} \frac{\partial}{\partial u} \sum_n e^{-nu} = -\frac{1}{Z} \frac{\partial Z}{\partial u} \qquad 13.11$$

where the last step uses the grand partition function from Equation 13.8. But by Equation 13.9, the grand partition function is also $Z = \left(1 - e^{-u}\right)^{-1}$. This can be used to find the derivative in Equation 13.11

$$\bar{n}_{BE} = -\frac{1}{Z} \frac{\partial Z}{\partial u} = -\left(1 - e^{-u}\right) \left[-\frac{e^{-u}}{\left(1 - e^{-u}\right)^2} \right]$$

$$= \frac{e^{-u}}{1 - e^{-u}} = \frac{1}{e^u - 1}$$

In terms of physical quantities, the result is

BE distribution

$$\bar{n}_{BE} = \frac{1}{e^{(\varepsilon-\mu)/k_B T} - 1} \qquad 13.12$$

This result, the *BE distribution*, plays the same role for bosons as Equation 13.7 does for fermions. Remarkably, the forms of the two distributions differ by only the sign attached to the number 1 in the denominator. However, as you have already seen, fermions and bosons differ greatly with respect to how they occupy quantum states. The distinction will become clearer in Section 13.1.6.

13.1.6 MB Distribution

In the classical limit—for example, the ideal gases studied in Chapter 6—the average occupancy of states is quite low ($\ll 1$), for either fermions or bosons. In that case, there will be no "competition" for space in a single state among multiple particles. This requires that the exponential term be much greater than 1 in either distribution (Equation 13.7 or 13.12). Therefore, in this limit, both distributions reduce to

---

MB DISTRIBUTION

$$\bar{n}_{MB} = \frac{1}{e^{(\varepsilon-\mu)/k_BT}} = e^{-(\varepsilon-\mu)/k_BT} \qquad\qquad 13.13$$

Two points of interpretation are worth mentioning. First, one may think of the MB distribution as simply the product of a constant e^{μ/k_BT} and the familiar Boltzmann factor $e^{-\varepsilon/k_BT}$ from classical statistics. Second, the devolution of both quantum distributions into the classical one is consistent with Bohr's correspondence principle, which connects the quantum and classical regimes.

13.1.7 Comparison of the Three Distributions

The three distributions are graphed together in Figure 13.3, for the same chemical potential μ. Notice that the BE distribution has the highest occupancy and the FD distribution the lowest, consistent with the example and discussion in Section 13.1.2. The three distributions approach one another in the classical limit.

For scale, note that when $\varepsilon = \mu$, the distribution laws require that $\bar{n}_{FD} = 0.5$ and $\bar{n}_{MB} = 1.0$ exactly. Occupancy of a state by fermions is required to be ≤ 1 at any energy. In contrast, the number of bosons in a quantum state is unbounded.

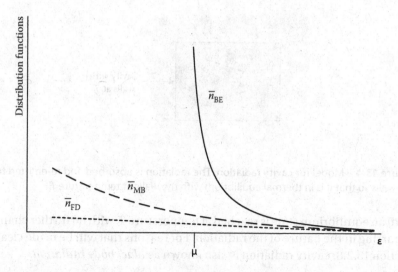

Figure 13.3 Comparison of the three distribution functions. At high energy, the two quantum distributions devolve into the classical (MB) distribution.

13.2 PHOTONS AND THERMAL RADIATION

The emission of electromagnetic radiation from hot bodies is an important example, with numerous applications throughout physics and astrophysics. In this section, we first treat the general subject and then apply BE statistics to find some results that can be compared with experimental observations.

13.2.1 Radiation Density and Pressure

It is an experimental fact that all bodies emit electromagnetic radiation by virtue of their temperature. This *thermal radiation* depends in general both on the temperature of the body and on the nature of its surface. A familiar example is a common electric heating element. At high temperatures, the element glows red or red-orange. Set on a lower temperature, the element emits primarily infrared radiation, which you can easily feel even though there is insufficient visible radiation to see. Bodies at higher temperatures, such as molten steel in a furnace, appear to glow white, indicating emission throughout the visible spectrum.

A useful model of thermal radiation is shown in Figure 13.4. In this context, it is called *cavity radiation*, because it is entirely contained within a cavity. The thermal radiation will be absorbed and re-emitted by the walls of the cavity

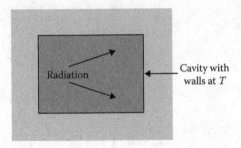

Figure 13.4 Model for cavity radiation. The radiation is absorbed and re-emitted from the walls so that it is in thermal equilibrium with the walls at temperature *T*.

until an equilibrium state is reached at temperature *T*, with no further changes occurring in the nature of the radiation. For reasons that will be made clear in Section 13.2.8, cavity radiation is also known as *blackbody radiation*.

It is possible to treat cavity radiation as a thermodynamic *PVT* system, such as those presented in Chapters 9 and 10, although at first sight it might seem an unlikely candidate. In this case, *V* and *T* are simply the cavity volume and temperature, respectively. To see how pressure arises, first recall that classical electromagnetism treats the radiation as electromagnetic waves. The energy density *u* of electromagnetic radiation is

$$u = \frac{B^2}{2\mu_0} + \frac{\varepsilon_0 E^2}{2} \qquad\qquad 13.14$$

for magnetic field *B* and electric field *E*. Therefore, the total energy *U* associated with the radiation is the energy density integrated over the entire volume of the cavity

$$U = \int_V \left(\frac{B^2}{2\mu_0} + \frac{\varepsilon_0 E^2}{2} \right) dV \qquad\qquad 13.15$$

There is a straightforward connection between the radiation pressure and energy density. Consider an electromagnetic wave incident on a boundary, as in Figure 13.5. The E_x field component of the wave induces a current density j_x in the wall. The B_y field component then interacts with j_x to produce a force F_z on the wall in the *z* direction. This is the origin of the radiation pressure.

To obtain the relation between *P* and *u*, start with the result from kinetic theory (Section 3.4) that the pressure in a gas is

$$P = \frac{1}{3} n m \overline{v^2} \qquad\qquad 13.16$$

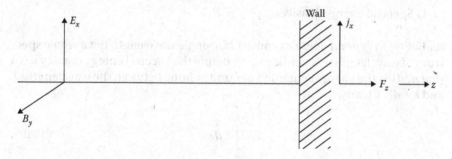

Figure 13.5   Radiation pressure arising from a reflected electromagnetic wave.

where n is the number density, m is the molecular mass, and $\overline{v^2}$ is the mean square molecular speed. (In this context m is not a particle mass but rather the mass-energy of the photons.) The product nm is the mass per unit volume or density ρ, so

$$P = \frac{1}{3}\rho\overline{v^2}$$ 13.17

Now consider the radiation as a *photon gas* where the photons are all moving with speed c. Equation 13.17 becomes

$$P = \frac{1}{3}\rho c^2$$ 13.18

According to the Einstein mass–energy relation $u = \rho c^2$, and so

$$P = \frac{1}{3}u$$ 13.19

This simple result expresses the connection between radiation pressure within the cavity and energy density.

> Do not confuse the radiation pressure in Equation 13.19 with the pressure exerted by light in other physical situations. For example, an external light beam exerts a force on a reflecting mirror due to photon momentum $p = h/\lambda$. As seen in the derivation, the result in this section is due to reflections from the cavity walls.

13.2.2 Energy Density for Cavity Radiation

In addition to the energy density u, there are several other important quantities related to cavity radiation:

1. Spectral energy density u_λ.

Radiation in general does not consist of a single wavelength but a whole spectrum of wavelengths. Accordingly, we define the spectral energy density u_λ so that $u_\lambda\, d\lambda$ is the energy contained per unit volume between the wavelengths λ and $\lambda + d\lambda$. Clearly,

$$u = \int_0^\infty u_\lambda d\lambda \qquad\qquad 13.20$$

2. Energy density per unit energy u_ε.

Similarly, u_ε is defined as the energy density within a range ε to $\varepsilon + d\varepsilon$.

3. Spectral absorptivity of a surface α_λ.

α_λ is defined as the fraction of energy incident on a surface that is absorbed at λ:

4. Spectral emissivity of a surface ε_λ.

Spectral emissivity is defined such that $\varepsilon_\lambda\, d\lambda$ is the energy emitted per unit area per second by the surface between λ and $\lambda + d\lambda$.

Figure 13.6 shows a box, with the walls at temperature T, in which there is cavity radiation. Experimentally, the radiation emitted by the exterior of the box to the outside depends on both the temperature of the walls and the nature of the walls. For example, a red smooth box looks quite different from a blue rough box. However, there is something special about the radiation within the box. The following statement about this radiation will be proven in Section 13.2.3:

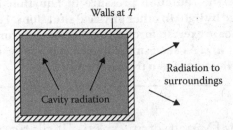

Figure 13.6   Radiation within a cavity depends on the temperature of the walls only. Radiation to the outside depends on the nature of the walls as well as their temperature.

Whatever the nature of the materials of the wall, the energy density for cavity radiation depends only on the temperature of the walls, while the spectral energy density depends only on the temperature and the wavelength.

That is,

$$u = u(T) \tag{13.21}$$

$$u_\lambda = u_\lambda(\lambda, T) \tag{13.22}$$

The spectral energy curves for cavity radiation were measured experimentally in the 19th century, and this was followed by attempts to understand the results theoretically. A famous attempt was made by Rayleigh and Jeans (see discussion in the text by Thornton, Rex, and Hood 2021). They assumed quite sensibly that radiation inside the cavity could be modeled by standing waves. This assumption led to a theoretical distribution $u_\lambda = u_\lambda(\lambda, T)$ that worked quite well at longer wavelengths but failed at shorter wavelengths, because the number (and energy) of short-wavelength waves grows without bound. This so-called *ultraviolet catastrophe* demonstrated the need for a better theory.

In 1900, Max Planck showed that the correct energy density and spectral distribution could be obtained by assuming that electromagnetic radiation is quantized in energy packets of size hf, where f is the frequency of the radiation and h is the very small constant now known as *Planck's constant*, with SI value approximately 6.626×10^{-34} J·s. For electromagnetic radiation $f = c/\lambda$, so in terms of wavelength, the quantized energy is hc/λ.

To find the energy density, begin with the Rayleigh–Jeans model of standing waves in a box, like the one shown in Figure 13.6. For a box of length L, the allowed wavelengths are $\lambda = 2L/n$ (where n is an integer), so the allowed energies are

$$\varepsilon = \frac{hc}{\lambda} = \frac{hcn}{2L} \tag{13.23}$$

Scaling up to three dimensions, the result is the same, but now n represents the magnitude of a "vector" in number space with three "components" n_x, n_y, n_z representing the quantum numbers for the three dimensions.

The total energy is given by quantum statistics as the sum of all possible energies weighted by the BE distribution (Equation 13.12). The chemical potential is zero in this case, because in cavity radiation the photon number is not conserved, and there is essentially no cost to create or destroy photons. Therefore, the total energy is

$$U = 2 \sum_{\text{all space}} \varepsilon \bar{n}_{BE} = 2 \sum_{\text{all space}} \frac{\varepsilon}{e^{\varepsilon/k_B T} - 1} \qquad 13.24$$

where the sum over all space represents all possible combinations of quantum numbers n_x, n_y, n_z. The factor 2 inserted in Equation 13.24 is due to the two possible photon polarizations for each standing wave.

The sum is evaluated by converting it to an integral in spherical coordinates (n, θ, φ). Because only positive values of n_x, n_y, n_z are allowed, this restricts the integral to the first octant of number space, so the result of the θ and φ integrals is $(1/8) \times 4\pi = \pi/2$. Therefore,

$$U = \pi \int_0^\infty \frac{n^2 \varepsilon}{e^{\varepsilon/k_B T} - 1} \, dn \qquad 13.25$$

because ε and n are connected (Equation 13.23), it is useful to express this integral entirely in terms of ε. Making the substitutions $n^2 = (2L\varepsilon/hc)^2$ and $dn = (2L/hc) \, d\varepsilon$, the integral becomes

$$U = \frac{8\pi L^3}{h^3 c^3} \int_0^\infty \frac{\varepsilon^3}{e^{\varepsilon/k_B T} - 1} \, d\varepsilon \qquad 13.26$$

Because L^3 is the assumed cavity volume in this model and $u = U/V$,

$$u = \frac{8\pi}{h^3 c^3} \int_0^\infty \frac{\varepsilon^3}{e^{\varepsilon/k_B T} - 1} \, d\varepsilon \qquad 13.27$$

Evaluating the integral will yield the energy density. First, however, note that the integrand (with constants) represents the photon spectrum, expressed per unit energy. That is,

$$u_\varepsilon = \frac{8\pi}{h^3 c^3} \frac{\varepsilon^3}{e^{\varepsilon/k_B T} - 1} \qquad 13.28$$

To evaluate the integral in Equation 13.27, first use the substitution $x = \varepsilon/k_B T$. This leaves

$$u = \frac{8\pi (k_B T)^4}{h^3 c^3} \int_0^\infty \frac{x^3}{e^x - 1} \, dx$$

The remaining integral is a standard one (as found by Euler) and can be expressed as a product of a gamma function and the Riemann zeta function. The result of that evaluation is

$$u = \frac{8\pi^5 (k_B T)^4}{15 h^3 c^3} \qquad \qquad 13.29$$

Notice that the energy density depends only on temperature, as stated at the beginning of this section.

13.2.3 Experimental Determination and the Stefan–Boltzmann Law

As a way to observe experimentally the photon radiation expressed in Equation 13.29, suppose a small hole is placed in one side of the cavity shown in Figure 13.6. The radiation emitted should be proportional to the area of the hole and is measured as energy per unit area per unit time, designated $R(T)$.

Dimensional analysis suggests that energy density u (energy/volume) should be multiplied by the photon speed c (distance/time) in order to get $R(T)$ (energy/area/time). This is nearly true, but two additional correction factors are needed: 1/2 because only half the photons are traveling in the desired direction (say left/right), and an additional 1/2 for the directional component (given by a cosine function) averaged over the entire surface. Therefore, $R(T) = cu/4$, with result

$$R(T) = \frac{2\pi^5 (k_B T)^4}{15 h^3 c^2} \quad \text{(Stefan–Boltzmann law)} \qquad 13.30$$

Equation 13.30 is the famous *Stefan–Boltzmann law*. Customarily, the constants attached to T^4 are lumped together into a single constant σ, the Stefan–Boltzmann constant, with value in SI units

$$\sigma = \frac{2\pi^5 k_B^4}{15 h^3 c^2} = 5.67 \times 10^{-8} \, \text{W/(m}^2 \cdot \text{K}^4)$$

Thus, $R(T) = \sigma T^4$. This result (but without Planck's constant) was known experimentally before Planck's theory.

This relationship for energy escaping from a hole holds equally well for energy emitted from the surface of a body at temperature T. A model often used for such a body is called a *blackbody*, defined as one that absorbs all incident radiation, so $\alpha_\lambda = 1$ for all wavelengths. If the body is not black but instead has

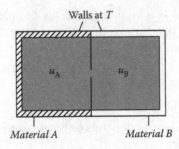

Figure 13.7 If the energy density of the cavity radiation in A is different from that in B, then the second law will be violated. For that reason, the energy density of cavity radiation depends on the temperature only.

a mean absorptivity α over all wavelengths with $\alpha < 1$, the Stefan–Boltzmann law has to be modified to give the energy radiated as $\alpha\sigma T^4$ rather than σT^4.

We are now in a position to return to the statements 13.21 and 13.22 and prove their validity. Consider the box shown in Figure 13.7. The box is composed of two halves, A and B, with the walls made of different materials but at the same temperature. Suppose that the energy density u in the two halves of the box is different, with say $u_A > u_B$. If the two halves are separated by a partition with a hole in it, energy will be incident on both sides of the hole, and there would be a net flux of energy from A to B, because $u_A > u_B$. Thus, B will heat and A will cool, with no external work being done on the system. This is a violation of the Clausius statement of the second law. A similar argument would hold if $u_B > u_A$. We conclude then that $u_A = u_B$. In other words, the energy density within a cavity depends only on the temperature of the walls and not on the nature of the walls, which is the statement 13.21.

This argument may now be extended to the spectral energy density u_λ simply by covering the hole with a filter passing radiation only in the narrow band between λ and $\lambda + d\lambda$. u_λ^A must then equal u_λ^B, by the argument just employed. From this, statement 13.22 follows.

13.2.4 Thermodynamic Derivation of the Stefan–Boltzmann Law

The energy equation for a PVT system is

$$\left(\frac{\partial U}{\partial V}\right)_T = T\left(\frac{\partial P}{\partial T}\right)_V - P \qquad 8.11$$

Also, the following relations have been shown to apply to cavity radiation:

$$P = \frac{1}{3}u, \; U = uV \text{ and } u = u(T)$$

Substituting these relations into Equation 8.11,

$$u = \frac{1}{3}T\frac{du}{dT} - \frac{1}{3}u$$

or

$$4\frac{dT}{T} = \frac{du}{u}$$

Integrating,

$$u = AT^4 \qquad\qquad 13.31$$

where A is a constant. This is consistent with the Stefan–Boltzmann law, as derived previously from quantum statistics.

13.2.5 Spectral Distribution

It is straightforward to change the energy spectrum distribution u_ε in Equation 13.28 to the spectral energy density u_λ. For photons $\varepsilon = hc/\lambda$, and therefore $d\varepsilon = (hc/\lambda^2)\,d\lambda$. (In taking this derivative, we ignore the resulting minus sign, which is irrelevant for the energy density.) Making these substitutions,

$$\frac{8\pi}{h^3c^3}\frac{\varepsilon^3}{e^{\varepsilon/k_BT}-1}d\varepsilon = \frac{8\pi hc}{\lambda^5}\frac{1}{e^{hc/\lambda k_BT}-1}d\lambda$$

Therefore, the spectral energy density is

$$u_\lambda = \frac{8\pi hc}{\lambda^5}\frac{1}{e^{hc/\lambda k_BT}-1} \qquad\qquad 13.32$$

Notice that u_λ is a function of both λ and T, consistent with Equation 13.21.

In practice, it is easier to measure the spectral emissivity ε_λ than the spectral energy density u_λ. Similar to the conversion process discussed in Section 13.2.3, this involves converting energy per unit volume per unit wavelength to energy per unit area per unit time, which requires the factor $c/4$. That is, $\varepsilon_\lambda = cu_\lambda/4$, with result

$$\varepsilon_\lambda = \frac{2\pi hc^2}{\lambda^5} \frac{1}{e^{hc/\lambda k_B T} - 1} \; (\text{Planck law}) \qquad 13.33$$

This result is usually referred to as the *Planck radiation law*. The quantity ε_λ gives the power per unit area per unit wavelength emitted from a cavity hole or from the surface of a perfect blackbody.

13.2.6 Analysis of Spectral Energy and the Wien Law

A graph of the spectral energy density is shown in Figure 13.8. This graph contains a wealth of information. First, notice how the curve changes as the temperature increases. The curve grows larger overall, indicating that more radiation is emitted as the temperature increases.

> **An alternate derivation of the Stefan–Boltzmann law involves integrating the spectral energy density or Planck law over all wavelengths.**

Notice also that each curve has a well-defined peak. The wavelength associated with this peak is defined as λ_{max}. Figure 13.8 shows that λ_{max} decreases as temperature increases. It makes physical sense that hotter objects should emit more photons with shorter wavelengths, because there is more energy available, and each photon's energy is hc/λ. The temperature dependence of λ_{max} is also consistent with the example of the electric heating element discussed earlier. A warm heating element emits almost entirely in the infrared part of the spectrum (longer wavelengths), but a much hotter element emits shorter-wavelength red and orange light.

In principle, it is straightforward to find the peak by setting $du_\lambda/d\lambda = 0$ and solving for $\lambda = \lambda_{max}$. This process is fairly tedious, so we leave it to the problems at the end of this chapter (Problem 13.16). The result is a simple expression

$$\lambda_{max} T = 2.898 \times 10^{-3} \, \text{m} \cdot \text{K} \; (\text{Wien law}) \qquad 13.34$$

which is known as the *Wien law* after physicist Wilhelm Wien, who deduced the law in 1896. Like the Stefan–Boltzmann law, the Wien law was known experimentally before Planck's theory.

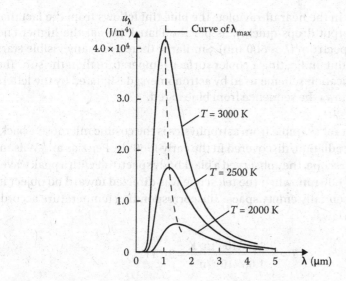

Figure 13.8 Spectral energy density as a function of λ and T for cavity radiation.

13.2.7 Some Applications

Given its general nature, blackbody radiation is fairly ubiquitous in scientific and everyday applications.

Observed spectra of stars follow closely to a theoretical blackbody, though with interruptions in the form of dark absorption lines corresponding to the star's elemental composition. Our sun is a good example. With surface temperature about 5780 K, the peak wavelength of the sun's radiation is given by the Wien law as

$$\lambda_{max} = \frac{2.898 \times 10^{-3}\,\text{m} \cdot \text{K}}{5780\,\text{K}} = 5.01 \times 10^{-7}\,\text{m} = 501\,\text{nm}$$

This is consistent with our observation of the sun as appearing fairly white, with emission distributed across the visible spectrum.

On the other hand, many stars exhibit a distinct bluish hue, indicating that they are hotter. One example is the bright nearby star Sirius A, which is easily visible to the naked eye and is in fact the brightest star in our sky. Its temperature is 9940 K, so its blackbody spectrum peaks at

$$\lambda_{max} = \frac{2.898 \times 10^{-3}\,\text{m} \cdot \text{K}}{9940\,\text{K}} = 2.92 \times 10^{-7}\,\text{m} = 292\,\text{nm}$$

This is actually in the near ultraviolet! The blue tint follows from the fact that the spectral output drops quickly, as per the Planck law, at the higher end of the visible spectrum ($\lambda > 500$ nm). Similarly, there are many visible stars with a reddish tint, indicating a cooler surface temperature than the sun. The spectral classification scheme used by astronomers (designated by the letters OBAFGKM) follows the sequence from blue to red.

Another important example from astrophysics is the cosmic microwave background (CMB) radiation, discovered in the early 1960s by Penzias and Wilson. With a radio telescope, they observed a blackbody spectrum with a peak wavelength close to 1.06 mm when the telescope was directed toward no object in particular—essentially empty space. The corresponding temperature according to the Wien law is

$$T = \frac{2.898 \times 10^{-3}\,\text{m} \cdot \text{K}}{1.06 \times 10^{-3}\,\text{m}} = 2.73\,\text{K}$$

The CMB provides, along with Hubble's law, solid evidence for the origin of the universe in a Big Bang approximately 13.7 billion years ago. It is evidence of a hot early universe, but with the radiation now severely redshifted to an equivalent temperature of 2.73 K.

Although the Big Bang is now well established, the CMB is still the subject of intense study. Two space-based telescopes—the Cosmic Background Explorer (COBE) in the 1990s and the Wilkinson Microwave Anisotropy Probe (WMAP) in the early 2000s—have carefully mapped the CMB throughout the sky. Both missions found strong correlation with a Planck blackbody curve. However, the latter probe was more sensitive and discovered some anisotropy in the radiation, which might have significant consequences for our understanding of the Big Bang or cosmology.

Back on Earth, humans and other warm-blooded animals act as blackbody radiators, as we are generally warmer than our surroundings. Wien's law predicts for a human body temperature of 310 K (37°C) a peak wavelength of 9.35 μm, squarely in the infrared part of the spectrum. Infrared thermometers, used to detect radiation and thereby determine temperature, are now in widespread use by medical practitioners and in industry. Radiative loss of energy from the body occurs (along with rapid cooling) when, for example, the head or hands are uncovered in extremely cold conditions. The Stefan–Boltzmann law, with radiative power proportional to T^4, explains why this is so important. You may have seen marathon runners given shiny foil capes at the end of a race, to wrap around themselves and prevent excessive radiative loss. The same radiative loss makes humans observable in the dark, through the use of infrared photography.

EXAMPLE 13.1

An electric stove heating element with a surface area of 80 cm² radiates energy at a rate of 420 W. (a) Find the temperature of the heating element. (b) What is the peak wavelength of the radiation?

Solution: (a) Using the Stefan–Boltzmann law (Equation 13.30), the emitted power is related to the temperature by $P = \sigma T^4 A$. This can be solved for the temperature: $T = (P/\sigma A)^{1/4}$. The numerical result is

$$T^4 = \left(\frac{P}{\sigma A}\right)^{1/4} = \left(\frac{420 \text{ W}}{\left[5.67 \times 10^{-8} \text{ W}/\left(m^2 \cdot K^4\right)\right]\left(0.0080 \text{ m}^2\right)}\right)^{1/4} = 981 \text{ K}.$$

This is equal to 708°C, a reasonable result for an electric burner on a high setting.

(b) The peak wavelength comes from the Wien law:

$$\lambda_{max} = \frac{2.898 \times 10^{-3} \text{ m} \cdot K}{T} = \frac{2.898 \times 10^{-3} \text{ m} \cdot K}{981 \text{ K}} = 2.95 \text{ μm}$$

The peak is in the infrared part of the spectrum. However, for a burner at this temperature, a significant part of the spectral distribution (Figure 13.8) falls below 700 nm (=0.7 μm). There may be a visible red-orange glow.

13.2.8 Kirchhoff Law

The Kirchhoff law is often quoted as "Good absorbers are good emitters." A more precise formulation is $\varepsilon_\lambda / \alpha_\lambda$ and is a constant for all surfaces, at a given temperature and wavelength.

The Kirchhoff law follows from our previous analysis for cavity and blackbody radiation. Refer again to Figure 13.6. If a body (not necessarily black) is placed inside the cavity, the radiation will be preserved within the cavity if the energy *absorbed* by the body per second, between the wavelengths λ and $\lambda + d\lambda$, is equal to the energy *radiated* between those wavelengths. That is,

$$\alpha_\lambda \frac{1}{4} c u_\lambda d\lambda = \varepsilon_\lambda d\lambda \qquad\qquad 13.35$$

The factor $c/4$ was explained in Section 13.2.5, and an additional factor α_λ now applies for imperfect blackbodies, with $\alpha_\lambda = 1$ only in the case of a perfect blackbody. Therefore, by Equation 13.22,

$$\frac{\varepsilon_\lambda}{\alpha_\lambda} = \frac{c}{4} u_\lambda(\lambda, T)\qquad 13.36$$

The right side of Equation 13.36 is a universal function of λ and T and is independent of the nature of the body. This implies that, at a given wavelength and temperature,

$$\varepsilon_\lambda = C\alpha_\lambda \qquad 13.37$$

where C is a constant that is the same for all bodies. This is the Kirchhoff law, as presented above.

Finally, there is one more important point. If the body is black, with $\alpha^{black} = 1$, Equation 13.36 gives

$$\varepsilon_\lambda^{black} = \frac{1}{4} c u_\lambda(\lambda, T)$$

This means that u_λ and $\varepsilon_\lambda^{black}$ have exactly the same dependence on λ and T. This explains the equivalence between cavity radiation and blackbody radiation.

13.3 APPLICATION OF FD STATISTICS TO ELECTRONS IN METALS

In the late 19th century, attempts were made to understand the properties of metals by modeling the free electrons in a metal as a classical ideal gas. Although electrical conduction can be modeled successfully this way, it is easily shown that this method fails to predict the electrons' contribution to the heat capacity. A gas of electrons, free to move in three dimensions, should have three degrees of freedom and therefore should contribute $3R/2$ to the molar heat capacity of a good conductor, such as copper. However, the observed heat capacity of copper is close to $3R$, as one would find considering only lattice vibrations (Section 3.4.4). There is a small contribution (on the order of $0.01R$) from the electrons, and this contribution increases with increasing temperature. It is evident that quantum statistics must be considered to solve the heat capacity problem.

13.3.1 Fermions at $T = 0$

Electrons are fermions, so the number of electrons per state is either 0 or 1, as per the FD distribution $\bar{n}_{FD}$. At $T = 0$, $\bar{n}_{FD}$ is a step function, as shown in Figure 13.2. It will also be useful to know the *fermion number density* $n(\varepsilon)$, defined as the number of particles per unit energy from ε to $\varepsilon + d\varepsilon$.

Notice that $n(\varepsilon)$ is analogous to the function u_ε, defined for bosons in Section 13.2.2.

The energy density $n(\varepsilon)$, and distribution function $\bar{n}_{FD}$ are related through the *density of states* $g(\varepsilon)$, defined as the number of states per unit energy from ε to $\varepsilon + d\varepsilon$. By their definitions, the three functions are related by

$$n(\varepsilon) = g(\varepsilon)\bar{n}_{FD} \qquad 13.38$$

Finding the density of states is the key step in understanding the distribution of energies for the conduction electrons.

To model the electrons in a metal, consider the quantized energy levels of identical particles of mass m trapped in a cube of side L. Solving the Schrödinger equation for this system leads to quantized energy levels

$$\varepsilon = \frac{h^2}{8mL^2}\left(n_x^2 + n_y^2 + n_z^2\right) = \frac{h^2 r^2}{8mL^2} \qquad 13.39$$

where the n_i are integer quantum numbers 1, 2, 3, ... and $r^2 = n_x^2 + n_y^2 + n_z^2$ gives the "radius" r in a number space (n_x, n_y, n_z), corresponding to energy ε, as shown in Figure 13.9.

At $T = 0$, the highest filled energy level corresponds to the chemical potential μ in Figure 13.2, and in this context, the highest energy is defined as the *Fermi energy* ε_F. This occurs at the largest number radius $r = R$, so

$$\varepsilon_F = \frac{h^2 R^2}{8mL^2} \qquad 13.40$$

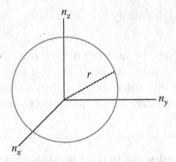

Figure 13.9 Number space (n_x, n_y, n_z) corresponding to energy ε in Equation 13.39. The "radius" $r = R$ corresponds to the maximum values of the quantum numbers.

That largest radius R (in number space) contains the total number of particles N, which is analogous to the "volume" of the "sphere" of radius R, with two important corrections:

1. The number of states is actually twice the number assumed to this point, due to the fermion spin degeneracy.

2. Only positive quantum numbers are allowed, so only one-eighth of the number sphere may be occupied.

Thus, $N = 2 \times \dfrac{1}{8} \times \dfrac{4}{3} \pi R^3$, or

$$N = \frac{1}{3} \pi R^3 \qquad\qquad 13.41$$

Combining Equations 13.40 and 13.41,

$$\varepsilon_F = \frac{h^2}{8m} \left(\frac{3}{\pi} \frac{N}{V} \right)^{2/3} \qquad\qquad 13.42$$

where we have used $V = L^3$ as the box's volume. Equation 13.42 is useful, because it allows computation of the Fermi energy for a real system with particle density N/V. For most metals, the Fermi energy is in the range 1–10 eV.

To find the density of states, consider a state with energy ε corresponding to radius r in the number sphere, with the relationship between ε and r given in Equation 13.39. Analogous to Equation 13.41, the number N_r of electrons up to radius r is $N_r = (1/3)\pi r^3$. Then, by Equation 13.39,

$$\varepsilon = \frac{h^2}{8mL^2}\left(\frac{3N_r}{\pi}\right)^{2/3}$$ 13.43

In terms of the Fermi energy (Equation 13.42),

$$\varepsilon = \varepsilon_F\left(\frac{N_r}{N}\right)^{2/3}$$ 13.44

Rearranging,

$$N_r = N\left(\frac{\varepsilon}{\varepsilon_F}\right)^{3/2}$$

By definition, the density of states is $g(E) = dN_r/d\varepsilon$, which reduces to

$$g(\varepsilon) = \frac{3}{2}N\frac{\varepsilon^{1/2}}{\varepsilon_F^{3/2}}$$ 13.45

Recall that the fermion number density is given by Equation 13.38. At $T = 0$, the FD distribution is a step function (Figure 13.2), so at this temperature

$$n(\varepsilon) = \begin{cases} g(\varepsilon) & \text{for } \varepsilon < \varepsilon_F \\ 0 & \text{for } \varepsilon > \varepsilon_F \end{cases} \quad (T=0)$$ 13.46

The result is shown in Figure 13.10(a). All levels are occupied up to the Fermi energy, but above the Fermi energy $n(\varepsilon) = 0$.

The distribution in Equation 13.46 can be used, for example, to compute the mean energy of the electrons in the distribution

$$\bar{E} = \frac{1}{N}\int_0^\infty \varepsilon n(\varepsilon)d\varepsilon = \int_0^{\varepsilon_F} \varepsilon g(\varepsilon)d\varepsilon$$

where Equation 13.46 was used for $n(\varepsilon)$. Now using Equation 13.45,

$$\bar{E} = \frac{3}{2\varepsilon_F^{3/2}}\int_0^{\varepsilon_F} \varepsilon^{3/2}d\varepsilon$$

which is easily evaluated to give

$$\bar{E} = \frac{3}{5}\varepsilon_F$$ 13.47

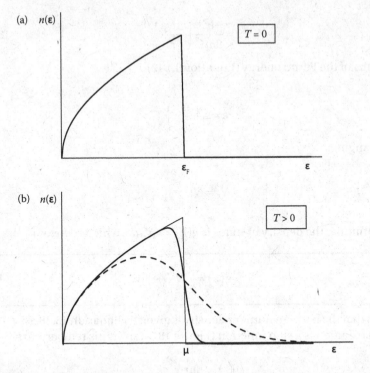

Figure 13.10   (a) Fermion number density function $n(\varepsilon)$ at $T = 0$. All levels are occupied up to the Fermi energy ε_F. (b) Fermion number density function $n(\varepsilon)$ at $T > 0$ (solid line) and $T \gg 0$ (dashed line). For reference, the $T = 0$ graph is included is a thin line. Notice that at high temperatures, the curve approaches the classical MB distribution.

This result seems quite realistic, given the shape of the distribution in Figure 13.10(a). For a collection of N fermions, the total energy is then

$$U = N\bar{E} = \frac{3}{5}N\varepsilon_F \qquad\qquad 13.48$$

Note that the Fermi energy ε_F and chemical potential μ are only equal at $T = 0$. The Fermi energy is simply the energy of the highest occupied state when $T = 0$, whereas μ for the system varies as a function of temperature.

EXAMPLE 13.2

Silver has a single 5s electron that gives it the property of having exactly one conduction electron per atom. Use this fact to estimate the Fermi energy of silver.

Solution: The Fermi energy can be computed using the known properties of silver with Equation 13.42. With one conduction electron per atom, the electron density N/V is the same as the number density of atoms. Standard handbook values for silver give a density of 10.49 g/cm³ and atomic mass 107.9 u. Therefore the number density N/V is

$$\frac{N}{V} = \left(1.049 \times 10^4 \, \text{kg/m}^3\right)\left(\frac{1 \, \text{atom}}{(107.9 \, \text{u})(1.661 \times 10^{-27} \, \text{kg/u})}\right) = 5.853 \times 10^{28} \, \text{m}^{-3}.$$

Now from Equation 13.42 the Fermi energy is

$$\varepsilon_F = \frac{h^2}{8m}\left(\frac{3}{\pi}\frac{N}{V}\right)^{2/3} = \frac{\left(6.626 \times 10^{-34} \, \text{J} \cdot \text{s}\right)^2}{8\left(9.11 \times 10^{-31} \, \text{kg}\right)}\left(\frac{3}{\pi} \times 5.853 \times 10^{28} \, \text{m}^{-3}\right)^{2/3} = 8.806 \times 10^{-19} \, \text{J}.$$

This is equal to 5.50 eV, which matches tabulated values.

13.3.2 Fermions at $T > 0$

When thermal energy is added to the system, some of the fermions absorb energy and end up with $\varepsilon > \varepsilon_F$. By Equation 13.38, the fermion number density is $n(\varepsilon) = g(\varepsilon)\bar{n}_{FD}$, or

$$n(\varepsilon) = \frac{g(\varepsilon)}{e^{(\varepsilon - \mu)/k_B T} + 1} \qquad\qquad 13.49$$

Figure 13.10(b) shows two distributions: one with the temperature somewhat larger than zero, and the other representing a temperature much larger than zero. Naturally, as the temperature increases, more fermions occupy higher energy states, and the distribution is "smeared" to higher energies. The distribution approaches a classical one at extremely high temperatures.

The situation is complicated somewhat by the fact that the chemical potential μ in Equation 13.49 is temperature dependent. In the $T = 0$ case (Section 13.3.1), it was safe to assume that $\mu = \varepsilon_F$, but this is no longer the case for $T > 0$. Rather, μ decreases as temperature increases. To see why this is so, note that μ is the energy at which the probability of a state being occupied is exactly 0.5. Since $g(\varepsilon)$ is an increasing function of ε, then μ must be smaller to compensate.

In this general case, the number of fermions is

$$N = \int_0^\infty n(\varepsilon)\,d\varepsilon = \int_0^\infty \frac{g(\varepsilon)}{e^{(\varepsilon-\mu)/k_B T}+1}\,d\varepsilon \qquad\qquad 13.50$$

and the total energy is

$$U = \int_0^\infty \varepsilon n(\varepsilon)\,d\varepsilon = \int_0^\infty \frac{\varepsilon g(\varepsilon)}{e^{(\varepsilon-\mu)/k_B T}+1}\,d\varepsilon \qquad\qquad 13.51$$

13.3.3 Application to Electron-Specific Heat Capacity

It is now possible to understand why electrons do not contribute $3R/2$ to a metal's specific heat, as they would do classically (Section 13.3 introduction).

First, note that in a typical metal μ (or ε_F) is on the order of 1–10 eV. This is much larger than thermal energy $k_B T$, except when the temperature is very high. (For example, $k_B T \approx 1/40$ eV at room temperature.) Now consider the shape of the $T = 0$ distribution in Figure 13.10(a). For most electrons in the distribution, absorbing $k_B T$ worth of energy is impossible, because they would end up in a fully occupied state. The only exception to this is electrons that are at the top of the distribution, initially less than $k_B T$ away from ε_F. Thus, the fraction of electrons subject to thermal promotion is something like

$$A\frac{k_B T}{\varepsilon_F}$$

where A is a constant slightly larger than 1, due to the shape of the distribution. In a detailed calculation, Sommerfeld found that $A = \pi^2/4$ in the limit of low temperatures. Thus, the thermal energy added to a collection of N electrons is approximately

$$\Delta U = N k_B T\left(\frac{\pi^2 k_B T}{4\varepsilon_F}\right) = \frac{\pi^2 N k_B^2 T^2}{4\varepsilon_F}$$

Using Equation 13.48, the total energy is now

$$U = \frac{3}{5}N\varepsilon_F + \frac{\pi^2 N k_B^2 T^2}{4\varepsilon_F} \qquad\qquad 13.52$$

The electronic contribution to the heat capacity (Equation 3.6) is $C_V = \partial U/\partial T$, or

$$C_V = \frac{\pi^2 N k_B^2 T}{2\varepsilon_F} \qquad\qquad 13.53$$

At room temperature (293 K), numerical evaluation gives $c_V \approx 0.02R$ per mole of electrons. This is in agreement with the measured result. Further, as the temperature is raised or lowered from room temperature, the electronic heat capacity varies as predicted by Equation 13.53. In the limit as the temperature approaches absolute zero, the heat capacity also approaches zero, as required by the third law of thermodynamics (Chapter 12).

13.4 BE CONDENSATION

An important application of current interest is *BE condensation*. The graph of the BE distribution in Figure 13.3 suggests that at sufficiently low temperature, a collection of identical bosons might lie mostly or entirely in the lowest energy state, given that there is no statistical limitation on how many bosons might occupy a single state. Such a state was first suggested by Bose and Einstein in 1924 (Bose 1924) and has since been observed in multiple ways.

13.4.1 Theoretical Model

We present here a theoretical model for estimating the maximum temperature for a particular BEC. More detailed models exist throughout the literature.

This model uses the same three-dimensional gas model as used for fermions in Section 13.3. The only adjustment needed is the factor of two between the boson and fermion distributions, due to the Pauli exclusion principle. Therefore, the fermion density of states in Equation 13.45 may be modified for bosons to

$$g_{BE}(\varepsilon) = \frac{2\pi V}{h^3}(2m)^{3/2}\,\varepsilon^{1/2} \qquad\qquad 13.54$$

In this context, m is the mass of each of the identical, indistinguishable bosons in the collection. The number distribution $n(\varepsilon)$, similar to Equation 13.38 for fermions, is

$$n_{BE}(\varepsilon) = g_{BE}(\varepsilon)\bar{n}_{BE}$$

$$= \frac{2\pi V}{h^3}(2m)^{3/2}\varepsilon^{1/2}\frac{1}{e^{(\varepsilon-\mu)/k_B T}-1}$$

13.55

where $\bar{n}_{BE}$ is given by Equation 13.12.

The theoretical model can be related to experiments. To begin, the number N of bosons in the collection is equal to the number distribution integrated over all energies. That is,

$$N = \int_0^\infty \frac{2\pi V}{h^3}(2m)^{3/2}\frac{\varepsilon^{1/2}}{e^{(\varepsilon-\mu)/k_B T}-1}d\varepsilon$$

or

$$N = \frac{2\pi V}{h^3}(2m)^{3/2}\int_0^\infty \frac{\varepsilon^{1/2}}{e^{(\varepsilon-\mu)/k_B T}-1}d\varepsilon$$

13.56

The BEC forms when the chemical potential is close to zero (Problem 13.32), so to a good approximation

$$N = \frac{2\pi V}{h^3}(2m)^{3/2}\int_0^\infty \frac{\varepsilon^{1/2}}{e^{\varepsilon/k_B T}-1}d\varepsilon$$

or

$$N = \frac{2\pi V}{h^3}(2mk_B T)^{3/2}\int_0^\infty \frac{x^{1/2}}{e^x-1}dx$$

13.57

after substituting $x = \varepsilon/k_B T$.

The value of the integral in Equation 13.57 is

$$\int_0^\infty \frac{x^{1/2}}{e^x-1}dx = \Gamma\left(\frac{3}{2}\right)\zeta\left(\frac{3}{2}\right)$$

$$\approx \frac{\sqrt{\pi}}{2}(2.61238\ldots) \approx 2.315$$

after approximating the Riemann zeta function. Therefore, the approximate answer to Equation 13.57 is

$$N = \frac{2\pi V}{h^3}(2mk_BT)^{3/2}(2.315)$$ 13.58

Equation 13.58 can be solved for temperature to yield T_c, called the *critical temperature*, which is the highest temperature at which the BEC should form. The result is

$$T_c = \frac{h^2}{2mk_B}\left[\frac{N}{2\pi V(2.315)}\right]^{2/3}$$ 13.59

The connection with experiments is possible because N/V in Equation 13.59 is simply the particle density. Thus, T_c can be estimated for a collection of bosons with given mass and density. For example, the number density of liquid helium at the point it reaches the superfluid stage is about 2.11×10^{28} m^{-3}. With a mass of 6.65×10^{-27} kg for a ^{4}He atom, this gives $T_c \approx 3.06$ K, which is less than 1 K above the observed transition temperature.

13.4.2 Experimental Observations

Superfluid behavior was first observed in ^{4}He in 1927 by Peter Kapitza and colleagues. Helium becomes liquid at 4.2 K at atmospheric pressure, but the temperature can easily be reduced further by reducing the pressure of the vapor surrounding the fluid. Once a temperature of 2.17 K is reached, superfluid effects are visible, and they are remarkable.

First, the name *superfluid* comes from the fact that the fluid can flow with essentially zero viscosity through even the smallest holes and capillaries. This low viscosity helps create a *creeping film* of fluid that can rise and flow over the walls of its container. Low viscosity also contributes to the *fountain effect* (or *thermomechanical effect*), where incident electromagnetic radiation causes the fluid to expand and thus rise through a small capillary and spray upward. As discussed in Section 10.5.2, the transition from normal to superfluid phases at 2.17 K is accompanied by a large spike in heat capacity. The transition temperature is often called the lambda point, because of the shape of the heat capacity graph (Figure 10.12).

Another interesting phenomenon that occurs below the lambda point is *second sound*, which is actually not sound but heat transfer that occurs in a wave-like pattern. The speed of second sound is not as high as the speed of normal sound waves; it is zero just below the lambda point and increases at lower temperatures, reaching a maximum of 20 m/s at 1.7 K (see Lane et al. 1947).

It is an important result of quantum statistics that the superfluid behavior of the rare isotope ^{3}He (a fermion) is vastly different than that of the boson ^{4}He. ^{3}He becomes a superfluid only at temperatures of 2.7 mK and lower and is aided by the presence of extreme pressure. It cannot be a true BEC, but instead it is thought that the fermions form pairs to create bosons, somewhat analogous to the electron-pairing mechanism in superconductors.

Strictly speaking, superfluids are not entirely equivalent to a BEC. First of all, the atoms in a superfluid are interacting. Second, only a fraction of the atoms in a superfluid are in the condensed state. However, this fraction approaches 1 as T approaches zero.

It was not until 1995 that Cornell and Wieman produced the first gaseous BEC, in a gas of rubidium atoms (the ^{87}Rb isotope) at very low pressure and temperature, about 170 nK. They used the technique of laser cooling to achieve such temperatures and an extremely dilute gas containing only a few thousand atoms. A BEC can exhibit unusual properties, such as quantized vortices (also found in superfluid helium). The vortices have been used to model the behavior of black holes. Another interesting effect observed is interference, attributed to the system's wave–particle duality.

Some further insight into the properties of the BEC may be gained by considering how its heat capacity varies with temperature. A typical graph of heat capacity as a function of temperature is shown in Figure 13.11. Notice that C_V approaches zero as the temperature approaches zero, as required by the third law (Chapter 12). Above the transition temperature T_c, the gas in the normal state has molar heat capacity $c_V = 3k_B/2$, consistent with the equipartition theorem. Below T_c, it can be shown (Problem 13.33) that the BE distribution gives internal energy U proportional to $T^{5/2}$. By Equation 3.6, the heat capacity $C_V = \partial U/\partial T$ is proportional to $T^{3/2}$, which matches the shape of the curve shown in Figure 13.11. Experimental data are consistent with the theoretical curve. The boundary between the condensed and normal states forms a continuous "cusp" that is easily seen, which is in contrast to the "spike" in the heat capacity at T_c for the superfluid transition (Figure 10.12). This is one of the clear distinctions between superfluids and BEC gases.

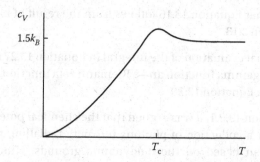

Figure 13.11   Molar heat capacity c_V as a function of temperature for a BEC.

Problems

13.1 Following the example of Section 13.1.2, find and count all the possible states for each of the three distributions (MB, FD, and BE) for the following cases: (a) two particles and three states and (b) three particles and two states. (c) Discuss any patterns you see developing, considering the example in this book and the two cases here in (a) and (b).

13.2 Following the example of Section 13.1.2, find and count all the possible states for each of the three distributions (MB, FD, and BE) for case of two particles and five states.

13.3 Verify that the sum of probabilities given by Equation 13.3 is exactly one.

13.4 For a single particle with energy ε and chemical potential μ, show that the average occupancy is given by

$$\bar{n} = -\frac{1}{Z}\frac{\partial Z}{\partial u}$$

where $u = (\varepsilon-\mu)/k_B T$.

13.5 Show that the formula derived in Problem 13.4 gives the correct distributions for both FD and BE statistics.

13.6 (a) Graph the FD distribution (Equation 13.7) using the following parameters: $\mu = 0.5\,\text{eV}$, $T = 300\,\text{K}$. (b) Expressed as a fraction of μ, how far from μ is ε when the distribution function is equal to (i) 0.9 and (ii) 0.1? (c) Repeat (a) and (b) for the same chemical potential but $T = 1000\,\text{K}$. (d) Discuss why the answers to (b) and (c) differ as they do.

13.7 Show that Equation 13.16 follows from the result of kinetic theory in Equation 3.13.

13.8 Show that evaluation of the integral in Equation 13.27 leads to a product of a gamma function and a Riemann zeta function, with the result given in Equation 13.29.

13.9 In Section 13.2.1, it was argued that the chemical potential should be zero for a collection of photons in cavity radiation. Make the same argument based on thermodynamic grounds. Hint: Consider the Helmholtz free energy.

13.10 (a) Calculate the temperature of the sun, assuming it to be a perfect blackbody, if the rate of solar energy reaching Earth's atmosphere is 1370 W/m² (called the solar constant). The radius of the sun is 6.96×10^8 m, and the mean distance of the sun from Earth is 1.496×10^{11} m, also known as one astronomical unit. (Slightly less than half the sun's radiation reaches Earth's surface, on average, but this still leaves over 600 W of power for each m² of Earth's surface. This explains the interest in solar panels.) (b) Find the power radiated per square meter on the sun's surface. (c) Find the total power generated by the sun and the corresponding rate of mass loss due to nuclear fusion.

13.11 Convert the spectral energy density u_λ to a frequency distribution u_f which expresses energy per unit volume per unit frequency, in terms of frequency f.

13.12 Express the Planck radiation law in terms of frequency f.

13.13 Use the Planck law to derive the Stefan–Boltzmann law by integrating ε_λ over all possible wavelengths.

13.14 Show that in the limit of long wavelengths, the Planck law may be approximated by

$$\varepsilon_\lambda = \frac{2\pi c k_B T}{\lambda^4}$$

(This is in fact the result obtained by Rayleigh and Jeans.)

13.15 Model a human body as a cylinder of height 1.70 m and radius 14 cm, with normal body temperature 37°C. Find the net blackbody radiation from the body per day, and compare with a normal daily food intake of 2000 kcal, if the environment is (a) room temperature 20°C; (b) a very cold day –10°C. (The results should convince you of the importance of clothing to insulate the body!)

13.16 Use the spectral energy density u_λ in Equation 13.32 to find the Wien law.

13.17 The tungsten filament of an old-style light bulb reaches a temperature of about 3300 K. Assume that it radiates as a perfect blackbody. (a) What is the peak wavelength λ_{max}? In what part of the spectrum does this lie? (b) What is the surface area of a 100-W filament? (c) Assess this device's efficiency as a light source by computing the fraction of the radiation that falls in the visible spectrum, 400–700 nm.

13.18 The red giant star Arcturus is one of the brightest visible stars. Its surface temperature is 4290 K. (a) Find the peak wavelength of radiation, assuming Arcturus is a perfect blackbody, and assess the result relative to the "red giant" label. (b) Arcturus is truly a giant, with a radius 1.77×10^{10} m, which is more than 25 times larger than the sun. Find the net power output from Arcturus and compare with that of the sun $(3.8 \times 10^{26}$ W$)$.

13.19 Show that MB statistics are valid for the common gas nitrogen at room temperature (20°C) and atmospheric pressure. Hint: Evaluate

$$N = \int_0^\infty n(\varepsilon)\,d\varepsilon$$

using the boson density of states (Equation 13.54) along with the MB factor in Equation 13.13. Thereby show that the constant factor e^{μ/k_BT} is small under the conditions given and use this fact to justify the use of MB statistics.

13.20 As a different approach to the preceding problem, consider the same gas but this time compute (a) the de Broglie wavelength (using the rms speed) and (b) the mean intermolecular spacing, assumed to be $V^{1/3}$, where V is the mean volume occupied by a single molecule in the gas. (c) Compare the results of (a) and (b) and discuss the implications for MB statistics.

13.21 Find the Fermi energy for copper, given a density of 8960 kg/m³ and exactly one conduction electron per atom.

13.22 Show that the thermal energy contained in the electrons in a metal is small by computing separately the first and second terms in Equation 13.52 at room temperature (293 K) for one mole of copper, which has $\varepsilon_F = 7.0$ eV.

13.23 The Fermi energy for aluminum is 11.7 eV. Use this to estimate the number of conduction electrons per atom in aluminum. Does your answer make sense?

13.24 Gold has a density of 19,300 kg/m³ and, like copper, has one conduction electron per atom. (a) Compute the Fermi energy for gold. (b) What would be the mean kinetic energy of the electrons in gold at 293 K if they were treated as a three-dimensional ideal gas? (c) Reconcile the vast difference between your answers in (a) and (b).

13.25 A neutron star is a close-packed collection of (almost entirely) neutrons, which are fermions. Compute the Fermi energy in a neutron star of typical size, mass 4.0×10^{30} kg and radius 11 km. Comment on the result relative to the Fermi energy of conduction electrons in a metal.

13.26 (a) Using Equation 7.4 from thermodynamics and the result that $U = N\bar{E} = (3/5)N\varepsilon_F$ (Equation 13.48), evaluate the pressure associated with conduction electrons in a metal, as a function of the Fermi energy and particle density N/V. The result is known as *degeneracy pressure* and is responsible for keeping the metal from collapsing to a larger density due to the attractive electrostatic forces. (b) Evaluate the degeneracy pressure numerically for copper, with Fermi energy 7.0 eV and density 8960 kg/m³.

13.27

(a) Show that the condition for BEC transition (Equation 13.59) can be written

$$n\lambda_{dB}^3 = 2.612$$

where $n = N/V$ is the particle density and λ_{dB} is the thermal de Broglie wavelength, given by

$$\lambda_{dB} = \sqrt{\frac{2\pi\hbar^2}{mk_BT}}$$

(b) Evaluate the thermal de Broglie wavelength for a typical BEC density of 10^{20} m^{-3} and discuss the implications.

13.28 Estimate the temperature at which liquid neon (density = 1200 kg/m³) should become a superfluid and use the result to explain why superfluid behavior is not found in neon.

13.29 For a typical BEC density of 10^{20} m^{-3}, estimate the maximum possible temperature for a BEC in ^{87}Rb and compare with the temperature 170 nK of the first observed BEC.

13.30 Find the number density required for a BEC to form in helium at room temperature (293 K). Use your result to analyze the likelihood of this happening.

13.31 Cornell and Wieman reported making a BEC with about 2000 ^{87}Rb atoms in a volume of 10^{-15} m^3. Estimate the maximum temperature of the BEC.

13.32 The text argues that $\mu \approx 0$ when a BEC forms. Justify this argument using the BE distribution and the fact that in a BEC essentially all the particles have condensed to the ground state.

13.33 Use the BE distribution to show that the internal energy of a BEC in the condensed state should vary with temperature as $T^{5/2}$. Thus, show that the heat capacity of a BEC should vary as $T^{3/2}$.

BIBLIOGRAPHY

Annett, J.F., *Superconductivity, Superfluids, and Condensates*, Oxford University Press, Oxford, 2004.

Bose, S.N., Plancks Gesetz und Lichtquantenhypothese, *Zeitschrift für Physik 26*, 178–181, 1924.

Cornell, E.A. and Wieman, C.E., The Bose-Einstein condensate, *Scientific American*, March, 40–45, 1998.

Davis, K.B., Bose–Einstein condensation in a gas of sodium atoms, *Physical Review Letters 75*(22), 3969–3973, 1995.

Lane, C.T., Fairbank, H.A., and Fairbank, W.M., Second sound in liquid helium II, *Physical Review 71*, 600–605, 1947.

Thornton, S.T., Rex, A., and Hood, C., *Modern Physics for Scientists and Engineers*, fifth edition, Cengage, Boston, MA, 2021.

Appendix A: Values of Physical Constants and Conversion Factors

Quantity	Symbol	Value
Gas constant	R	8.31 J/(K·mol)
Avogadro constant	N_A	6.02×10^{23} mol^{-1}
Boltzmann constant	k_B	1.38×10^{-23} J/K
Stefan–Boltzmann constant	σ	5.67×10^{-8} W/(m²·K⁴)
Planck constant	h	6.63×10^{-34} J·s
Elementary charge	e	1.60×10^{-19} C
Faraday constant	$F_0 = eN_A$	9.65×10^{4} C
Speed of light[a]	c	3.00×10^{8} m/s
Acceleration due to gravity	g	9.81 m/s²
Permeability of free space	μ_0	$4\pi \times 10^{-7}$ H/m
Permittivity of free space	ε_0	8.85×10^{-7} F/m
Mechanical equivalent of heat	J	4.19 J/cal
Molar volume of an ideal gas at STP	v	22.4 L
Atmospheric pressure		1.01×10^{5} N/m² = 101 kPa = 760 mm of Hg = 760 torr
1 horsepower	hp	746 W
1 kilowatt hour	kW·h	3.60×10^{6} J
Bohr magneton	μ_B	9.27×10^{-24} J/T

[a] In the modern SI system, the speed of light is defined to be the exact nine-digit number 299,792,458 m/s.

Appendix B: Some Mathematical Relations Used in Thermodynamics

B.1 RECIPROCAL AND CYCLICAL RELATIONS

Suppose that there exists a relation between the variables x, y, and z

$$F(x,y,z)=0 \qquad\qquad \text{B.1}$$

so that only two of them are independent. Equation B.1 can be rearranged to give x as a function of y and z:

$$x=x(y,z) \qquad\qquad \text{B.2}$$

where $x(y, z)$ stands as usual for a function of y and z. The infinitesimal change dx in x resulting from infinitesimal changes dy and dz in y and z is

$$dx=\left(\frac{\partial x}{\partial y}\right)_z dy+\left(\frac{\partial x}{\partial z}\right)_y dz \qquad\qquad \text{B.3}$$

Similarly, writing $y = y(x, z)$,

$$dy=\left(\frac{\partial y}{\partial x}\right)_z dx+\left(\frac{\partial y}{\partial z}\right)_x dz \qquad\qquad \text{B.4}$$

Now using dy from Equation B.4 in Equation B.3:

$$dx=\left(\frac{\partial x}{\partial y}\right)_z\left(\frac{\partial y}{\partial x}\right)_z dx+\left[\left(\frac{\partial x}{\partial y}\right)_z\left(\frac{\partial y}{\partial z}\right)_x+\left(\frac{\partial x}{\partial z}\right)_y\right]dz \qquad\qquad \text{B.5}$$

Arbitrarily, one can choose x and z to be the independent variables. This means that it is possible to have $dz = 0$ in Equation B.5 and still have a non-zero value for dx, because x and z are independent. With this value of dz and with the common term dx cancelled, Equation B.5 becomes

$$1 = \left(\frac{\partial x}{\partial y}\right)_z \left(\frac{\partial y}{\partial x}\right)_z$$

or

$$\boxed{\left(\frac{\partial x}{\partial y}\right)_z = \left(\frac{\partial y}{\partial x}\right)_z^{-1}}$$

B.6

This is known as the *reciprocal relation*.

Alternatively, one could choose $dx = 0$ with $dz \neq 0$ in Equation B.5, yielding

$$\left(\frac{\partial x}{\partial y}\right)_z \left(\frac{\partial y}{\partial z}\right)_x + \left(\frac{\partial x}{\partial z}\right)_y = 0$$

or

$$\boxed{\left(\frac{\partial x}{\partial y}\right)_z \left(\frac{\partial y}{\partial z}\right)_x \left(\frac{\partial z}{\partial x}\right)_y = -1}$$

B.7

using Equation B.6. This is known as the *cyclical relation* or simply the *cyclical rule*. It is easy to remember because of the cyclical order. Note the –1 on the right-hand side.

B.2 CHAIN RULE

Suppose again that x, y, and z are not independent, being related by Equation B.1. Consider some function ϕ of x, y, and z. Because of Equation B.1, ϕ may be expressed in terms of only two of the variables, say

$$\phi = \phi(x, y)$$

B.8

Equation B.8 can be rearranged to give

$$x = x(\phi, y)$$

B.9

so

$$dx = \left(\frac{\partial x}{\partial \phi}\right)_y d\phi + \left(\frac{\partial x}{\partial y}\right)_\phi dy$$

B.10

Dividing Equation B.10 through by dz, holding ϕ constant:

$$\boxed{\left(\frac{\partial x}{\partial z}\right)_\phi = \left(\frac{\partial x}{\partial y}\right)_\phi \left(\frac{\partial y}{\partial z}\right)_\phi}$$

B.11

This is the *chain rule*. Note the common ϕ outside each partial derivative. The chain rule must not be confused with the cyclical rule, which is a relation just between the variables x, y, and z with no other function ϕ involved.

B.3 THE CONDITION FOR A DIFFERENTIAL TO BE EXACT

A mathematical function $\phi(x, y)$ of x and y takes unique values for each pair of values of x and y. When x and y change by dx and dy, the infinitesimal change in ϕ is

$$d\phi = \left(\frac{\partial \phi}{\partial x}\right)_y dx + \left(\frac{\partial \phi}{\partial y}\right)_x dy$$

B.12

Because it is the differential of a mathematical function, $d\phi$ is called *an exact differential*. A finite change in ϕ when x changes from x_1 to x_2 and y from y_1 to y_2 is

$$\Delta\phi = \phi(x_2, y_2) - \phi(x_1, y_1) = \int_{x_1 y_1}^{x_2 y_2} d\phi$$

B.13

With the values of ϕ fixed at the points (x_1, y_1) and (x_2, y_2), then $\Delta\phi$ is also fixed; consequently, it does not matter how x and y may vary during the integration between the given limits. This means that the integral is *path independent*. In thermodynamics it is frequently important to know whether an integral is path independent; in other words, it is necessary to establish whether the integrand is an exact differential. There is a simple test for this.

Suppose there is a differential of the form

$$dG = Xdx + Ydy \qquad\qquad \text{B.14}$$

where X and Y are in general functions of both x and y. We wish to establish now whether dG is exact. Differentiating the coefficient of dx in Equation B.12 with respect to y, while holding x constant, gives $\partial^2\phi/\partial y\partial x$, while differentiating the coefficient of dy with respect to x, while holding y constant, gives $\partial^2\phi/\partial x\partial y$. It is shown in the standard texts on partial differentiation that these two partial derivatives are equal; in other words, the order of the differentiation is immaterial. If dG is an exact differential, then

$$\boxed{\left(\frac{\partial X}{\partial y}\right)_x = \left(\frac{\partial Y}{\partial x}\right)_y} \qquad\qquad \text{B.15}$$

This argument shows that Equation B.15 is a *necessary* condition for dG to be exact; it may also be shown, using a more sophisticated argument, to be sufficient.

As an example, consider the differential

$$dG = 2xy^4 dx + 4x^2 y^3 dy \qquad\qquad \text{B.16}$$

One may verify that Equation B.15 is satisfied by the differential dG given in Equation B.16. Therefore, dG is an exact differential. In fact, the actual function G is yielded immediately upon integration. From the form of dG in Equation B.16,

$$\left(\frac{\partial G}{\partial x}\right)_y = 2xy^4 \quad\text{so}\quad G(x,y) = x^2 y^4 + f(y)$$

$$\left(\frac{\partial G}{\partial y}\right)_x = 4x^2 y^3 \quad\text{so}\quad G(x,y) = x^2 y^4 + g(x)$$

The only way for these two solutions for G to be equal is for $f(y) = g(x) =$ a constant. Thus $G(x, y) = x^2y^4 +$ a constant. By similar means, it may be established that the differential

$$G = xy^4 dx + 4x^2y^3 dy$$

is inexact.

B.4 INTEGRALS USED IN STATISTICAL MECHANICS

In Chapter 6, a number of definite integrals involving the Gaussian factor of the form

$$\int_0^\infty x^n e^{-ax^2} dx \qquad \text{B.17}$$

(where a is some constant or combination of constants) arise and must be evaluated. The first such definite integral is encountered in Section 6.4.1 in the analysis of the distribution of velocities v_x in a one-dimensional gas:

$$Z = \int_{-\infty}^\infty e^{-mv_x^2/2k_BT} dv_x$$

To begin the evaluation, first note that the integrand is an even function, so that the definite integral over infinite limits can be replaced by two times the same integral with semi-infinite limits:

$$Z = 2\int_0^\infty e^{-mv_x^2/2k_BT} dv_x \qquad \text{B.18}$$

This is a useful step because tabulated integrals are generally given with limits of zero to infinity. This now has the standard Gaussian form:

$$\int_0^\infty e^{-ax^2} dx$$

where a is a constant. The result can be worked out in a delightful way through a transformation to polar coordinates, and the result is

$$\int_0^\infty e^{-ax^2}\,dx = \frac{1}{2}\sqrt{\frac{\pi}{a}} \qquad\qquad \text{B.19}$$

For the partition function in Equation B.18, the constant is $a = m/2k_\mathrm{B}T$, and therefore by Equation B.19

$$Z = \sqrt{\frac{2\pi k_\mathrm{B}T}{m}}$$

in agreement with the result given in Section 6.4.1.

The more general definite integral in Equation B.17 also has well-known results, which are tabulated in printed tables and now found in standard computational software and online tools. The results can be related to the gamma function, which is of great interest in mathematics apart from its usefulness in this context. For purely computational purposes it is easier to avoid the gamma function and express the result in two forms, one for even values of the exponent and another for odd values:

$$\int_0^\infty x^{2n} e^{-ax^2}\,dx = \frac{1\cdot3\cdot5\cdot\ldots(\text{up through } 2n-1)}{2^{n+1}a^n}\sqrt{\frac{\pi}{a}} \qquad\qquad \text{B.20}$$

and

$$\int_0^\infty x^{2n+1} e^{-ax^2}\,dx = \frac{n!}{2a^{n+1}} \qquad\qquad \text{B.21}$$

As an example, you can easily verify that Equation 6.14 follows from application of Equation B.20, and Equation 6.17 follows from application of Equation B.21.

Appendix C: The Work Required to Magnetize a Magnetic Material and to Polarize a Dielectric

C.1 MAGNETIC WORK

Consider a sample of magnetic material becoming magnetized by being placed inside a long solenoid, as in Figure C.1. The length of the sample is l, the cross-sectional area is A, and it fits exactly inside the whole volume of the solenoid. The current is quasistatically increased, and with it the applied magnetization is also increased.

The following relations hold:

1. $B = \mu_0(H + M) = B_0 + \mu_0 M$

where B is the magnetic field, H is the auxiliary field, M is the magnetization or the magnetic moment per unit volume, B_0 is the magnetic field in the absence of the specimen, and μ_0 is the permeability of free space.

2. $M = \chi_m H$

for a linear magnetizable material such a paramagnet, where χ_m is the magnetic susceptibility.

3. $B = \mu_0(1 + \chi_m)H = \mu\mu_0 H$

where $\mu = 1 + \chi_m$ is the permeability.

4. $\mathcal{M} = MV$

where $\mathcal{M}$ is the overall magnetic moment.

For simplicity, it is assumed that all the vector quantities in the above relations are parallel, and they may be treated as scalars. Also, the vector quantities are

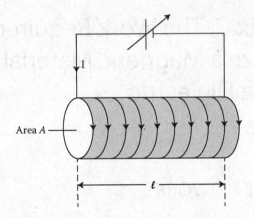

Figure C.1 A magnetic material contained within a solenoid. The current is gradually increased from zero so that the material is magnetized.

considered uniform over the volume V of the long solenoid, so that any end effects are ignored.

From basic electromagnetism, the auxiliary field in the middle of a long solenoid is $H = nI$, where I is the current and with n is the number of turns per unit length. This means that

$$B_0 = \mu_0 nI$$

The flux threading the solenoid is

$$\Phi = BAnl = BnV$$

If the current is increased from its instantaneous value I to $I + dI$ in a time dt, there is a back EMF

$$\varepsilon = nV\frac{dB}{dt}$$

It is the battery driving charge around the circuit against this back EMF that is the source of the work required to magnetize the sample.

The work done by the battery in the time dt, when charge Idt flows, is

$$W = \varepsilon Idt = nV\frac{dB}{dt}Idt = \frac{B_0 V}{\mu_0}dB$$

or

$$W = \frac{B_0 V}{\mu_0}\left[dB_0 + \mu_0 dM\right]$$

Then the total work done in the magnetization process is

$$W = V\int\frac{B_0 dB_0}{\mu_0} + V\int B_0 dM = V\int\frac{B_0 dB_0}{\mu_0} + \int B_0 d\mathcal{M}$$

There are two terms here. The first is just the familiar energy term that the solenoid would have in the absence of the magnetic sample; upon integration, it gives the familiar energy density $(B_0{}^2)/(2\mu_0)$. The second is the work required to bring the sample up to its final magnetization. We conclude that the infinitesimal work required to increase the overall magnetic moment from $\mathcal{M}$ to $\mathcal{M} + d\mathcal{M}$ in the applied field B_0 is

$$\boxed{W = B_0\, d\mathcal{M}}$$

If the magnetization and the magnetic field are not constant over the volume of the sample as we have assumed, this argument may be extended to give

$$đW = \int B_0 dMdV$$

where the integration takes place over the whole volume of the sample.

C.2 DIELECTRIC WORK

Now consider polarizing a dielectric quasistatically by placing it between the plates of a parallel plate capacitor as in Figure C.2, and gradually increasing the potential difference υ across the plates. Assume that the dielectric exactly fills the space between the plates and that the electric field is uniform, so that edge effects can be neglected.

The following relations hold:

1. $D = \varepsilon_0 E + P$

2. $P = \varepsilon_0 \chi_e E$

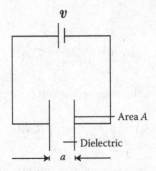

Figure C.2   A dielectric material between the plates of a parallel plate capacitor. The voltage across the capacitor is gradually increased from zero so that the material is polarized.

3. $D = \varepsilon_0(1 + \chi_e)E = \varepsilon_0\varepsilon E$

4. $P = PV$

where E is the electric field, D electric displacement, P polarization or the electric dipole moment per unit volume, ε_0 permittivity of free space, $\varepsilon = 1 + \chi_e$ relative permittivity, χ_e electric susceptibility, and p overall electric dipole moment. As for the magnetic case, assume that all the vector quantities in the above relations are parallel, and so the variables may be treated as scalars.

If the uniform free charge density on the plates is σ, Gauss's law gives

$$D = \sigma = \frac{Z}{A}$$

where Z is the charge on each plate. Now increase the charge by dZ. The battery has to do work

$$W = \upsilon dZ = (aE)(AdD) = VEdD$$

$$= V(\varepsilon_0 dE + dP)E$$

where the volume is $V = aA$. Integrating,

$$W = \varepsilon_0 V\int EdE + V\int EdP$$

The first term is the familiar energy term for the energy of an empty charged capacitor with an energy density $\varepsilon_0 E^2/2$ between the plates. The second term is then the work done in polarizing the dielectric. Therefore, the infinitesimal work required to increase the overall dipole moment of a dielectric from p to $p + dp$ in the field of E is

$$\mathrm{d}W = Edp$$

If the polarization and the electric field are not constant over the volume of the sample as we have assumed, this argument may be extended to give

$$W = \int EdPdV$$

where the integration is over the volume of the dielectric.

Appendix D: Answers to Selected Problems

CHAPTER 1

3. 327.79°; 6.83 cm; no.

4. 341.79°.

5. 348.35 K; 75.20°C.

6.

Liquid/Solid	Copper Nickel Thermocouple		Platinum Resistance Thermometer		Constant-Volume H_2 Thermometer		Constant-Volume H_2 Thermometer	
	E (mV)	T_E (°)	R (O)	T_R (°)	P (atm)	T_P (°)	P (atm)	T_P (°)
N_2	−0.10	−9	1.96	54	1.82	73	0.29	79
O_2	0.00	0	2.50	69	2.13	86	0.33	90
H_2O	5.30	486	13.65	379	9.30	374	1.37	374
Sn	9.20	827	18.56	516	12.70	510	1.85	505
At T.P.	2.98	273	9.83	273	6.80	273	1.00	273

The accepted (gas thermometer) values are 77.4 K (N_2), 90.2 K (O_2), 373 K (H_2O), and 505 K (Sn). The second constant-volume thermometer is as good as or better than the first in all cases, and in some cases it matches the handbook value to the nearest degree. This is not surprising, because the best results should be obtained at lower gas pressures, as per Equation 1.5.

CHAPTER 2

1. (a) 5.7×10^4 J. (b) -5.7×10^4 J.

5. (a) $(P_2 - P_1)(V_2 - V_1)$. (b) -2.02×10^4 J.

9. -92 J.

12. (a) 4.3×10^5 N. (b) 9.0 m.

13. $W = F\int_{T_1}^{T_2} \alpha L\, dT.$

14. 0.38 J.

15. 9.0×10^7 N/m²; no difference.

CHAPTER 3

1. (a) No. (b) Yes. (c) ΔU positive and same as (b).

2. (a) No. (b) No. (c) It increases. Remember that, in $\Delta U = Q + W$, we exclude any changes in the bulk KE and PE (see Section 3.2.1). Here the total energy, U + the bulk (PE + KE), remains constant, as there is no input of energy in the form of heat or work, and the organized motion of the rotational kinetic energy is converted into the random motion of internal energy (hence raising the temperature).

3. (a) No. (b) Yes, Q negative. (c) Negative.

4. (a) $Q = 60$ J. (b) $Q = -70$ J. (c) 50 J and 10 J absorbed.

5. (a) 0.10 m/s. (b) Work. (c) Heat.

6. $W = -3.46 \times 10^3$ J; $Q = 3.46 \times 10^3$ J; $\Delta U = 0$.

15. 7.6×10^5 J.

16. (a) 2.41×10^{-3} J/(K·mol); 0.302 J/(K·mol). (b) 7.53 J/mol. (c) 9.41×10^{-3} J/(K·mol).

18. 696 m.

20. 208 years.

22. (a) $P(V_2 - V_1)$. (b) $nRT\ln\left(\dfrac{V_2-nb}{V_1-nb}\right)+n^2a\left(1/V_2-1/V_1\right)$.

24. 664 kJ.

25. (a) -4.80×10^5 J/kg. (b) -4.99×10^5 J/kg.

CHAPTER 4

1. (a) 0.45. (b) 550 kJ/min.

2. No, the state of battery changes in the process as its stored energy changes.

4. No; although energy is conserved ($W = Q_1 - Q_2$), the reported efficiency η = 0.55 is larger than the Carnot efficiency η_C = 0.50.

5. Lowering temperature of cold reservoir, because it has a larger effect on the temperature ratio.

8. 1.2×10^6 J.

9. 1.6 kW.

15. $\eta = 0.38$.

CHAPTER 5

1. -1460 J/K.

4. (a) -60.6 J/K. (b) -13.1 J/K. (c) -12.2 J/K.

5. 0.226 J/K.

6. 0.424 J/K; 424 J/K.

7. (a) 67°C. (b) 0.60 J/K; 0.60 J/K.

9. 16 J/K.

CHAPTER 6

8. (a) $I = 6.3 \times 10^{-45}$ kg·m^2. (b) $E_0 = 3.8 \times 10^{-23}$ J $= 2.4 \times 10^{-4}$ eV; $Z = 106$ which is much greater than 1, so the approximation is valid.

10. (a) 1.37×10^9. (b) ≈ 1.0; 5.6×10^{-7}; 6.7×10^{-8}. (c) The partition function has changed significantly, but the probabilities are the same.

12. (a) 2.02. (b) 0.14 eV. (c) $P(0\ \text{eV}) = 0.49$; $P(0.2\ \text{eV}) = 0.31$; $P(0.4\ \text{eV}) = 0.20$.

14. (a) 17 mm/s. (b) 0.54 mm/s. (c) 35%.

15. (a) 0.6 m/s. (b) 1.0 m/s.

21. (a) 146 J/K. (b) 176 J/K.

CHAPTER 7

2. $P = RT/(v - b) - a/v^2$. This is the van der Waals equation.

3. 1.16×10^6 J.

4. $Pv = RT + BP + CP^2 + DP^3$ for one mole.

7. $\Delta g = -8.43 \times 10^5$ J/mol; 2.3×10^4 J/mol given out (exothermic).

11. (b) It is not related! Equation 7.49 gives $W_{\text{useful}} = -\Delta G$ for those processes in which the end points are at (P_0, T_0). This is not so for this problem.

12. (a) -237 kJ/mol. (b) 1.23 V.

CHAPTER 8

10. (a) Reversible; U, H. (b) Reversible; none. (c) Reversible; S. (d) Irreversible; U, H. (e) Irreversible; H.

13. (c) $-V\beta T\left(P_2 - P_1\right) + V\left(P_2^2 - P_1^2\right)/2B$.

15. -2.3 K.

17. (a) 0. (b) 3.72×10^6 J. (c) 1.15×10^4 J/K.

CHAPTER 10

1. (a) The liquid level gradually goes down until only vapor is left at Z. (b) The liquid–vapor interface becomes blurred as the critical point is approached, with the vapor and the liquid becoming indistinguishable.

2. (a) 31.9 Torr. (b) 17.5 Torr. (c) 1.1×10^{-7} g. Neglect volume of water in capillary.

5. 3.8×10^4 J/mol.

6. (a) 200 K, 1.01 atm. (c) $2R$ (per mol).

7. $S_{solid} > S_{liquid}$.

8. –5.8 K.

9. 0.05 K.

CHAPTER 11

9. (b) 2.6×10^{10} Pa.

10. $\mu_{NH_3} = \dfrac{1}{2}\left(\mu_{N_2} + 3\mu_{H_2}\right)$.

CHAPTER 12

1. No.

3. For a boson gas all the particles can be in the lowest possible state, so the temperature should be lower for bosons. One might expect the energy to be lower than the Fermi energy for electrons, which is the energy of the highest occupied state at $T = 0$.

CHAPTER 13

1. (a) MB 9, FD 3, BE 6. (b) MB 8, FD not possible, BE 4.

2. MB 25, FD 10, BE 15.

10. (a) 5780 K. (b) 6.33×10^7 W/m². (c) 3.85×10^{26} W; 4.28×10^9 kg/s.

12. $\varepsilon_f = \dfrac{2\pi h f^3}{c^2} \dfrac{1}{e^{hf/k_B T} - 1}$.

17. (a) 878 nm (infrared). (b) 1.5×10^{-5} m². (c) 0.12.

21. 7.0 eV.

Index

Printed in the United States
by Baker & Taylor Publisher Services

Printed in the United States
by Baker & Taylor Publisher Services